Synthesis Lectures on Information Concepts, Retrieval, and Services

Series Editor

Gary Marchionini, School of Information and Library Science, The University of North
Carolina at Chapel Hill, Chapel Hill, NC, USA

This series publishes short books on topics pertaining to information science and applications of technology to information discovery, production, distribution, and management. Potential topics include: data models, indexing theory and algorithms, classification, information architecture, information economics, privacy and identity, scholarly communication, bibliometrics and webometrics, personal information management, human information behavior, digital libraries, archives and preservation, cultural informatics, information retrieval evaluation, data fusion, relevance feedback, recommendation systems, question answering, natural language processing for retrieval, text summarization, multimedia retrieval, multilingual retrieval, and exploratory search.

Reagan W. Moore

Trustworthy Communications and Complete Genealogies

Unifying Ancestries for a Genealogical History of the Modern World

Second Edition

Reagan W. Moore
School of Information and Library Science
University of North Carolina at Chapel Hill
Chapel Hill, NC, USA

ISSN 1947-945X   ISSN 1947-9468 (electronic)
Synthesis Lectures on Information Concepts, Retrieval, and Services
ISBN 978-3-031-16838-3   ISBN 978-3-031-16836-9 (eBook)
https://doi.org/10.1007/978-3-031-16836-9

This Springer imprint is published by the registered company Springer Nature Switzerland AG
The registered company address is: Gewerbestrasse 11, 6330 Cham, Switzerland

Preface

The second edition of *Trustworthy Communications and Complete Genealogies: Unifying Ancestries for a Genealogical History of the Modern World* defines a coherence metric for evaluating whether unifying ancestries are missing essential information needed to identify progenitors. A genealogy that connects to the Unifying Ancestry for a national community is complete. However, all members of the national community should be able to connect. The ancestors of any member could be chosen as the Unifying Ancestry. The coherence metric then provides a way to differentiate between possible Unifying Ancestries and select the best one.

An extended version of the Research Genealogy based on 330,610 persons is used. The Unifying Ancestry for Western Europeans is based on 28,124 common ancestors of the Kings of Spain, Great Britain, Belgium, the Netherlands, Norway, and Sweden and the Queen of Denmark.

The genealogy metric for coherence is extended to include both external relationships (links to history books) and internal relationships (analysis for whether essential information is missing). The latter metric is then used to develop an estimate of the number of generations needed to trace ancestry back to a time when any person alive is a common ancestor of all persons in the chosen population. A Unifying Ancestry that requires the smallest number of generations is the best one.

The proposed list of progenitors for persons of Western European descent has been modified to consider all ancestors who are lineage coalescence points for own cousin relationships. Multiple filters are then applied to select ten potential progenitors.

Example lineages of Maximal Ascent to Charlemagne are provided for 27 persons identified by both Hart *"The 100: A Ranking of the Most Influential Persons in History"* and Palmer *"A History of the Modern World"*.

Chapel Hill, USA   Reagan W. Moore

Contents

Introduction

Information Science provides a way to identify properties of genealogies, construct a Unifying Ancestry for a national community, and link the Unifying Ancestry to a Genealogical History of the Modern World. The challenges that need to be addressed include the extraction of information from multiple sources, development of algorithms that analyze genealogy properties, and demonstration that genealogies can be linked to historical events. We start with an analysis of the properties that are needed to trust information derived from multiple sources.

Communication is the exchange of information between a sender and a receiver. Communication is trustworthy if both the sender and receiver interpret the information within the communication using the same context. Context is defined by a knowledge base that organizes relationships between the information elements within the communications. If both participants share the same context for interpreting the information, the communication will succeed. If they do not share the same context, the interpretation by the receiver will differ from the intent of the sender, and the communication will fail. If the communications lack essential information, no useful conclusions can be drawn from the information exchange.

The communication context includes:

- the date that the communication was initiated,
- the source of the information contained within the communication,
- the structural relationships needed to parse information from the communication,
- the semantics used to interpret information elements, and
- a set of knowledge relationships that are defined between the information elements.

Each participant has an internal knowledge base that they use to package information for exchange. In addition, there usually is a community-consensus knowledge base that

R. W. Moore, *Trustworthy Communications and Complete Genealogies*, Synthesis Lectures on Information Concepts, Retrieval, and Services, https://doi.org/10.1007/978-3-031-16836-9_1

governs terms used in the domain of discourse. When both participants agree on a shared context and reference the same community-consensus knowledge base, information can be reliably exchanged.

How can communication be trustworthy when a shared context is not available? Two worthwhile cases are communication with the future, and interpretation of communication from the past. In both cases, a shared context may not be available. These two cases can be used to quantify properties that trustworthy communications should possess.

Communication with the Future

Preservation is communication with the future. Archivists preserve records that document historical events for access by future generations. The Preservation community has developed two international standards that describe the communication context:

- ISO 14721—the Open Archival Information System, (Consultative Committee for Space Data, 2012). OAIS defines a context for each record composed of provenance information, authenticity information, integrity information, description information, representation information, and identification of a knowledge community.
- ISO 16363:2012—Audit and certification of Trustworthy digital repositories. (Technical Committee: ISO/TC 20/SC 13 Space Data, 2012). ISO 16363 defines the information that is needed to track whether the required OAIS information is present, how well the Archives are being managed and whether the Archives are trustworthy.

Can the trustworthiness assessment criteria for Archives be used to define a context for trustworthy communications?

The preservation community effectively defines a context for interpreting the records that comprise the communication with the future. The context consists of:

- Representation information that defines how to parse the record to extract information.
- Provenance information that defines who created the record and who submitted the record to the preservation environment.
- Authenticity information that assigns a unique identifier to the record and tracks whether versions of the record have been created.
- Integrity information that defines whether the record has been corrupted.
- Description information that defines the meaning of the record.
- Specification of a knowledge community that provides an external knowledge base for interpreting relationships between records.
- Management information for tracking changes to the preservation environment.
- Usage information for tracking recipients of the records.
- Security information for tracking unauthorized access.

- Auditing information for tracking the internal operations of the preservation environment.

Given this context, a person in the future should be able to correctly interpret and use information contained in the preserved records. For trustworthy preservation, either the context needed for successful communication is explicitly created and stored with the associated records or procedures are provided for dynamically generating the information (Moore, 2016). The records together with the context and procedures are archived for use in the future. However, the external knowledge base is not archived. Preservation depends on the continued existence of a knowledge community that maintains the community-consensus knowledge base.

Communication from the Past

The context needed to interpret communication from the past may not be available. Instead, a context may need to be created based upon the information content derived from multiple communications. Effectively, a set of multiple communications is turned into a knowledge base by analyzing relationships between the information elements present within the communications. Relationships that can be established between the information elements can then be used to analyze the trustworthiness of the communications by comparing the derived knowledge base with a community-consensus knowledge base. Since every collection has properties related to Consistency, Correctness, Connectivity, Closure, Completeness, and Coherence, these properties can be used to evaluate the resulting context.

The steps needed to build a missing context for interpreting communications from the past are:

- Identify the information elements that will be extracted from the communications.
- Associate a list of sources with each information element.
- Identify the types of relationships that will be established between the information elements. Examples might be spatial relationships, or temporal relationships, or semantic relationships, or category membership.
- Build a graph database that uses the relationships to define links between the information elements.
- Identify a community-consensus knowledge base for the domain of discourse.
- Analyze the epistemological properties of collections related to Consistency, Correctness, Connectivity, Closure, Completeness, and Coherence. This will require defining metrics for evaluating each property that are relevant to the communication domain of discourse.

- Consistency measures whether all the attributes needed to interpret information elements are provided.
- Correctness measures whether information values fall within acceptable ranges.
- Closure measures whether information elements are isolated, disconnected from the rest of the information elements in the communications.
- Completeness measures whether all information elements can be linked to a unifying topic. The unifying topic describes the intent of the communications.
- Connectivity measures how the information elements may be grouped together.
- Coherence measures whether the derived collection can be integrated with a community-consensus knowledge base (through creation of external relationships) and whether the derived collection is missing essential information needed to draw conclusions about the unifying topic (through the analysis of internal relationships).
• Finally, construct bi-directional links between the information elements in the database and the information elements in the community-consensus knowledge base. Once this is done, it then becomes possible to interpret the information content present in the communications.

This procedure constructs a context comprised of the relationships between information elements extracted from multiple communications. The communications can be interpreted correctly when both the sender and receiver agree on how the information items are linked to a community-consensus knowledge base.

Genealogy Test Case

Genealogists interpret communications from the past. A practical example of the generation of a local knowledge base is the construction of a genealogy. Records that document historical events are parsed to extract information about marriage, birth, and death; residence, titles, and occupations; familial relationships; and education, religion, and cause of death. Genealogies are viewed as trustworthy if they extract information from authoritative sources. Primary records that document the historical events are usually considered to be authoritative. Primary record examples include birth certificates, marriage licenses, tombstones, etc.

Genealogists would like to extract information from authoritative sources, in the expectation that a trustworthy genealogy will then be created. Genealogists rely upon provenance information (sources), authenticity information (the expectation that the source has not been modified), descriptive information (type of historical events), and representation information (how to interpret dates and locations). The trustworthiness of the genealogy then relies strongly upon the trustworthiness of the sources from which the information has been extracted.

Note there is a strong synergy between the information that Archives require about records and the information that Genealogists parse. A genealogy can be viewed as an index into the records in an archive that identifies familial relationships between the persons involved in the historical events.

The preservation standards for trustworthiness were not published until 2012. Typical sources used for genealogies predate the development of the standards. Also, preservation of records is done independently of the accuracy of the information contained within the records. Genealogists need not only sources that are authoritative, but also sources that contain accurate information. Can metrics be defined that evaluate the accuracy of the information contained within the sources that represent communications from the past?

A genealogy is a collection of information about persons. Every genealogy has generic properties that include Consistency, Correctness, Closure, Connectivity, Completeness, and Coherence. For genealogies, we can evaluate:

- Consistency—identify the attributes that need to be extracted about each person and verify that each person in the genealogy has the standard set of attributes.
- Correctness—identify external constraints that the attributes should not violate. For example, we can verify that the ages at marriage, birth of a child, and death fall within accepted biological ranges.
- Closure—verify that each person in the genealogy has a connection to every other person in the genealogy.
- Connectivity—identify the ways in which persons may be grouped, including coalescence of lineages to common ancestors.
- Completeness—identify a unifying ancestry for all members of a national community. Analyze the unifying ancestry to identify progenitors for the national community.
- Coherence—link the familial relationships in the genealogy to a community consensus knowledge base of historical events. Persons involved in the historical events should also be linked to the genealogy. Finally, identify whether essential information is missing that is needed to identify progenitors.

The Completeness property can be interpreted as the inverse of relationships related to Consistency. In addition to verifying that all the members of a genealogy have a standard attribute such as a familial relationship, a set of progenitors are defined to which all members of the national community should be able to link their ancestry. By linking persons in the genealogy to these progenitors, the genealogy can be immediately integrated with other genealogies that also have lineages to the progenitors. Effectively, the genealogy is inverted from a focus on the ancestors of a root person to a focus on the descendants of a group of progenitors. The Unifying Ancestry is the link between a focus on ancestors of a root person and a focus on Progenitors that are ancestors of all members of a national community.

The property of Coherence can be viewed as the identification of links from the genealogy to an external community-consensus knowledge base. An example is identifying lineages from the genealogy to historically important persons identified in history books. Such linking assumes that it is possible to identify ancestors for historically notable persons, and then link the genealogy to the same ancestors. By linking the ancestors to historical events, a Genealogical History of the Modern World can be constructed. This, in turn, becomes a community knowledge base for genealogies.

The Coherence property also requires that the genealogy be analyzed to determine whether essential information needed to identify progenitors is missing. Does the genealogy have the required connectivity to enable all members of a national community to link their ancestry to a set of common progenitors? The coherence metric calculates how many generations into the past lineages need to be traced to find a progenitor. Genealogies that require fewer generations have a better representation of lineage coalescence.

Genealogies serve as a useful test case for the evaluation of collections of communications. Each communication (source) is parsed for information, the information is organized in a database (genealogy), and analyses are then performed upon the database to verify collection properties. Relationships between the information elements are used to construct a local knowledge base, such as the Progenitors of a Unifying Ancestry. The local knowledge base can then be used to analyze the trustworthiness of the genealogy by linking the local knowledge base to external knowledge bases, in this case history books.

Unifying Ancestries

Genealogies possess a fundamental symmetry that is driven by two competing processes, the doubling of the number of potential ancestors each generation, and the exponential growth of lineage coalescence. Typically, by the sixth generation, lineages will start to coalesce. Multiple ancestors of the root person of the genealogy will have the same parents. By 1325 AD, the number of potential ancestors each generation exceeds the population of Europe. Lineages must then coalesce because there are not enough persons to fill all the potential ancestors for that generation. The degree of coalescence increases the further the genealogy is traced into the past.

The result is a symmetric genealogy. The root of the genealogy has a lineage to the two parents, and the number of lineages doubles each generation. A maximum number of ancestors found per generation in the Unifying Ancestry typically occurs in the thirteenth century. Because of the resulting high degree of coalescence, each ancestor will have many ascents, unique paths through the genealogy that terminate at the same person. By selecting those persons with the largest number of ascents, a set of progenitors for the genealogy can be defined. The genealogy can then be characterized by a graph that starts with a link from a root person, grows to a large number of ancestors, and then contracts back to a small number of progenitors.

The descendants of the progenitors constitute a unifying ancestry. A research question is whether this unifying ancestry will be the same for all members of a national community. Can a set of progenitors be found that comprise the common ancestors of all members of a national community such as persons of Western European descent? One measure of completeness for a genealogy is to demonstrate that a person's ancestors are linked to the progenitors for the national community. The unifying ancestry represents a knowledge base whose properties can be analyzed and linked to community-consensus knowledge bases such as history books.

Every national community has a Unifying Ancestry to which all members of the national group should be able to link their ancestry. A genealogy can be considered complete when relationships can be defined to all other members of the relevant national community. A Unifying Ancestry simplifies this task, since once you have a connection to your community's Unifying Ancestry, it is the responsibility of the other members of your national community to make their own connections.

A Unifying Ancestry is viable if connections can be made using events within recorded history. This means that a Unifying Ancestry needs to be based on historical fact, rather than being an artefact of cultural tradition. A Unifying Ancestry needs to have a high degree of connectivity through marriages between ancestral lineages. A connection to the Unifying Ancestry should directly lead to familial relationships with the other members of the Unifying Ancestry. Finally, a Unifying Ancestry should contain historically notable ancestors for each national community.

For all genealogies there is a correlation between availability of information about ancestors, and the number of ancestors that can be found. When lineages are traced far enough into the past, typically the only persons that can be identified are historically notable persons. A Unifying Ancestry should include information about historically notable persons who are members of the national community and provide lineages to these historically notable persons.

The existence of a Unifying Ancestry is driven by biological, historical, and social factors:

- Biologically, the number of potential ancestors doubles every ancestral generation. By 1325, the potential number of ancestors in a single generation exceeds the available population of Europe. A Unifying Ancestry for Western Europeans could then be created by creating a genealogy based on all the persons alive in 1325 in Europe. Fortunately, we can create a much smaller Unifying Ancestry.
- Historically, the amount of information available about persons decreases as you go back in time. Eventually, the only persons for which we can find information are historically notable persons. This suggests that a Unifying Ancestry will connect to historically important persons, such as Royal Families.
- Socially, the children of historically notable persons tend to marry the children of historically notable persons. This is particularly true for the Royal Families of Western

Europe, the Kings of Belgium, Great Britain, the Netherlands, Norway, Spain, and Sweden and the Queen of Denmark. A genealogy based on the common ancestors of these Royal Families is a strong candidate for a Unifying Ancestry for persons of Western European descent and will include historical events that comprise a history of the modern world. The lineages for the Royal Families comprise the most heavily researched lineages and therefore will be the most trustworthy.

A question that must be addressed is whether the descendants of Kings and Queens will include commoners, or whether the descendants will only include Royalty. An extensive analysis of the descendants of King George I, Elector of Hanover, was published in "The Book of Kings: A Royal Genealogy" (McNaughton, 1973). McNaughton developed an authoritative genealogy for the descendants through a 20-year exchange of personal communications with the Earl Mountbatten of Burma and Mountbatten's relatives. McNaughton identified 2962 descendants with birth locations in 51 countries, ranging from Europe to North and South America to Eastern Europe, the Middle East, and the Far East. The descendants included titled nobility, clergy, military, doctors, and commoners. The descendants of King George I provide an illustrative example of the number of generations that descents must be traced to find non-royal descendants. Typically, younger siblings become members of the peerage. In the 5th generation from George I, Gustav Karl von Reichenbach-Lessonitz is the Count of Reichenbach-Lessonitz. When younger siblings emigrate to other countries, their children tend to marry commoners. In the 9th generation from King George I, Maximilian von Pagenhardt was born in the United States and married a commoner. The expectation is that lineages linking commoners to Royal Families will require tracing ancestry back at least 7–10 generations.

The design of a Unifying Ancestry (genealogy knowledge base) raises multiple issues:

- How can lineage coalescence be measured?
- How does lineage coalescence lead to a Unifying Ancestry?
- Can a Unifying Ancestry encompass multiple national communities?
- Can a Unifying Ancestry be based on common ancestors of Royal Families?
- How many generations of descendants must be traced from Kings and Queens to find non-royal descendants?
- How far back in time do lineages need to be traced to connect to a Unifying Ancestry?
- How does the number of generations to a progenitor depend upon the size of the Unifying Ancestry?
- Does the connection with a Unifying Ancestry occur within historic times or is the connection based on cultural traditions?
- How can the required connectivity be measured?
- What fraction of your ancestry comes from the Unifying Ancestry?
- For what fraction of the members of the national community do we expect to have enough information to link their ancestry to the Unifying Ancestry?

- How many generations do lineages need to be traced to link to progenitors for your national community?
- Can the persons involved in historically important events be linked to the Unifying Ancestry?
- Can a Unifying Ancestry be used to create a Genealogical History of the Modern World?

References

Consultative Committee for Space Data. (2012). *Reference model for an open archival information system (OAIS): Recommended practice issue 2.* ISO 14721:2012.

McNaughton, A. (1973). *The book of kings: A royal genealogy.* Gladstone Press.

Moore, R. W., Xu, H., Conway, M., Rajasekar, A., Crabtree, J. & Tibbo, H. (2016). *Trustworthy policies for distributed repositories.* Morgan & Claypool

Technical Committee: ISO/TC 20/SC 13 Space Data. (2012). *ISO 16363:2012 Audit and certification of trustworthy digital repositories.* International Standards Organization.

Research Genealogy

To answer these questions, a 330,610 person Research Genealogy has been assembled and analyzed using the CoreGen3 Genealogy Analysis Workbench (Moore, 2020). Versions of the Research Genealogy have been published on MyHeritage (https://www.myheri tage.com/site-383545952/moore-family-history) and Geneanet (https://gw.geneanet.org/ rwmoore_w). The Research Genealogy is bundled with the CoreGen3 program and is available as free application software from the Microsoft Store (Windows application) and the Mac App Store (Cocoa application). More information about CoreGen3 and the Research Genealogy is available at https://www.coregen.center. The genealogy workbench was developed in tandem with the Research Genealogy to explore properties related to local and global connectivity, identification of Unifying Ancestries, identification of Progenitors for persons of Western European descent, and evaluation of the traditional completeness metrics of a royal descent or a descent from Charlemagne. All results that are presented in this book can be reproduced using the published database and genealogy analysis workbench.

To demonstrate that the Completeness property for a genealogy is consistent with finding a link to a Unifying Ancestry, we need to show that the Research Genealogy itself is trustworthy. The trustworthiness of the Research Genealogy is evaluated by analyzing its consistency, correctness, closure, connectivity, completeness, and coherence. The connectivity analysis is used to find a set of common ancestors. The common ancestors are then reduced to a set of progenitors. Finally, the progenitors are linked to an external knowledge base consisting of the historically notable persons listed in Palmer's History of the Modern World (Palmer & Colton, 1952).

This process identifies a set of algorithms for evaluating properties of genealogies (listed in Appendix E). The algorithms can be applied to authoritative genealogies, provided that:

R. W. Moore, *Trustworthy Communications and Complete Genealogies*, Synthesis Lectures on Information Concepts, Retrieval, and Services, https://doi.org/10.1007/978-3-031-16836-9_2

- The authoritative genealogies trace ancestries sufficiently far into the past.
- The authoritative genealogies are large enough to track interactions between multiple communities and the resulting lineage coalescence.
- The authoritative genealogies provide lineages to persons involved in historical events.

A goal for a genealogist is to build the unifying ancestry to which any member of a national community should be able to link their ancestry, and then link the unifying ancestry to the history of the modern world. For collections of communications, a similar analysis can be done on the information content to build a communication knowledge base, which is then linked to community-consensus knowledge bases that describe the physical world.

Research Genealogy Creation

A testbed for evaluating the properties of Unifying Ancestries was constructed by adding 50 persons every night for 27 years to the Research Genealogy. In parallel, a genealogy analysis workbench was developed over a 15-year period to analyze the trustworthiness of the genealogy. Each time a new analysis property was added to the workbench, the Research Genealogy was updated to provide the required information. Each time major extensions were made to the Research Genealogy new analyses were added to the workbench.

The construction of the Research Genealogy proceeded in multiple stages, from the identification of the ancestors of a single royal house, to the identification of multiple families that were linked to the royal house, to the identification of intermarriage between Royal Families, to the identification of common ancestors of the Royal Families, to the identification of Progenitors for the common ancestors, and then to integration with European history.

The initial motive was to verify a family tradition that we were 23rd cousins of Queen Elizabeth. My grandmother joined the society "Daughters of the American Revolution" and developed a genealogy with lineages back to William Wentworth in Lynn, Massachusetts in the 1630s. The Wentworth family published a two-volume genealogy on their ancestors in England and their descendants in the United States (Wentworth, 1878). This provided lineages back to William Marbury, Henry de Percy (known as Harry Hotspur,) and William the Conqueror. The Royal Lineage to Queen Elizabeth was then added to the Research Genealogy from lineages published in "Collins's Peerage of England" (Bridges, 1970).

McNaughton identified 16 Noble Families that married descendants of King George I, Elector of Hanover. For each family, a progenitor could be identified and all members of the Research Genealogy with the same family name were then linked to their progenitor through the lineages published in Europaische Stammtafeln (Schwinnicke, 1995).

This was possible because Europaische Stammtafeln lists all the children for each family. This led to a goal of identifying a progenitor for each Noble House and creating extended families linking all family members to their progenitor.

Within the Research Genealogy, lineages for the Royal Families of Europe were added. Table 1 lists ancestral information for the Kings of Spain, Great Britain, Belgium, the Netherlands, Norway, and Sweden and the Queen of Denmark. The ancestral information includes the relationship of each King and Queen to King Philippe of Belgium (Rel), the number of ancestors (#Anc), and the number of ancestors who are also members of the Unifying Ancestry (#Core Ancestors). The Unifying Ancestry is based on the ancestors of Prince George of Cambridge (the great grandson of Queen Elizabeth of Great Britain). This Unifying Ancestry automatically includes all the common ancestors of the Royal Families of Western Europe.

The number of common ancestors of the Royal Families present in the Research Genealogy is 28,124 persons. These common ancestors are the basis for the creation of a Unifying Ancestry for persons of Western European descent based on the ancestors of a descendant of a current King or Queen. In the Research Genealogy, each Royal Family has at least 28,600 ancestors with King Charles III having the largest number of ancestors (37,855). Since the Unifying Ancestry is based on the ancestors of his grandson, King Charles III has 37,855 ancestors who are members of the Unifying Ancestry. King Willem of the Netherlands has the smallest number of ancestors who are members of the Unifying Ancestry at 28,276 ancestors. Note that the relationship notation 2C 1R means 2nd cousin, once removed. Appendix A describes how to interpret cousin relationships. The current Kings and Queens range from 1st cousin to 4th cousin of King Philippe of Belgium.

Table 1 Royal Families of Europe

	Person	Rel	#Anc	#CoreAnc
1	Philippe Leopold Louis Marie of Belgium	Source	35,427	32,349
2	Margrethe II of Denmark	2C 1R	29,037	28,620
3	Charles Philip Mountbatten-Windsor	3C 1R	37,855	37,855
4	Willem Alexander of the Netherlands	4C 1R*	28,972	28,276
5	Harald V of Norway	1C 1R	28,652	28,511
6	Felipe VI of Spain	4C	31,725	29,315
7	Carl XVI Gustav of Sweden	3C	30,151	28,569

The unifying ancestry is the ancestors of George Alexander Louis Mountbatten-Windsor
Rel is the relationship to the first person in the list
#Anc is the number of ancestors of the person
#CoreAnc is the number of ancestors of the person in the unifying ancestry
* means that the number of generations from the source to the common ancestor is less than the number of generations from the person to the common ancestor

The ability to find a progenitor for a family was extended to find progenitors for a group of persons. Debrett published a chart for the ancestry of the current Kings and Queens of Western Europe that identified a common ancestor, Louise Auguste of Mecklenburg-Strelitz (Williamson, 1991).

Can a set of royal progenitors be found for the current Kings and Queens of Western Europe? An analysis was added to CoreGen3 to do this comprised of the following steps.

- Find the ancestors of each current King or Queen
- Generate the set of common ancestors
- Find the most recent common ancestor and mark the person as a royal progenitor
- Delete the ancestors of the marked person and the marked person from the list
- Iterate until no one is left in the list.

The CoreGen3 analysis of the common ancestors of the Royal Families produces a set of 31 royal progenitors (shown in Table 2) who along with their ancestors comprise the common ancestors of the Kings and Queens of Western Europe.

In Table 2, the column with a D denotes persons identified in Debrett. The column with an M denotes persons who were identified by McNaughton. The column with an S denotes persons identified in Europaische Stammtafeln. The column labeled ID denotes the person identifier used in the Research Genealogy.

There are multiple noteworthy facts in Table 2.

- The most recent royal progenitor is Louise Auguste of Mecklenburg-Strelitz, born in 1776.
- The lineages in the Debrett chart included six of the royal progenitors listed in Table 2 while the lineages in McNaughton included 19 of the royal progenitors.
- The lineages published in Europaische Stammtafeln include all 31 of the royal progenitors.
- The person ID was incremented by "1" each time a new person was added to the Research Genealogy. Therefore, Katharine zu Solms-Rodelheim (born in 1702) was the last royal progenitor added to the Research Genealogy given her person ID of 99,748. Increasing the size of the Research Genealogy to 330,610 persons did not lead to additional royal progenitors.
- All the royal progenitors are members of German Noble houses, with the exception of Frederick III of Denmark.
- The range of birth years for the royal progenitors is limited, varying from 1776 back to 1508. A possible explanation is the high number of intermarriages between the Royal Families during this period within the Habsburg Empire.
- Lineages to the royal progenitors of the Royal Families are based on historical fact.
- Lineages to the royal progenitors are provided by authoritative sources.

Table 2 Royal progenitors of the Royal Families

	Person	ID	Birth	Death			
1	Louise Auguste of Mecklenburg-Strelitz	24,686	1776	1810	D	M	S
2	Frederick William III of Prussia	24,656	1770	1840	D	M	S
3	Charlotte of Mecklenburg-Strelitz	89,846	1769	1818		M	S
4	Friedrich of Saxe-Altenburg	25,300	1763	1834			S
5	Wilhelmine Karolina of Orange-Nassau	25,134	1743	1787	D	M	S
6	Friederike Sophie of Brandenburg-Schwedt	25,106	1736	1798	D	M	S
7	Karl Christian of Nassau-Weilburg	25,119	1735	1788	D	M	S
8	Friedrich II Eugen of Wurttemberg	24,970	1732	1797	D	M	S
9	Charlotte Sophia of Saxe-Coburg-Saalfeld	25,236	1731	1810		M	S
10	Ludwig of Mecklenburg-Schwerin	25,235	1725	1778		M	S
11	Augusta of Saxe-Gotha	4206	1719	1772		M	S
12	Frederick Lewis of Hanover	4205	1707	1751		M	S
13	Dorothea Renata von Zinzendorf	94,212	1669	1743			S
14	Christine of Mecklenburg-Gustrow	93,537	1663	1749		M	S
15	Philipp of Hesse-Philippsthal	12,145	1655	1721		M	S
16	Katharina Amalie of Solms-Laubach	12,146	1654	1736			S
17	Ludwig Christian I of Stolberg-Gedern	91,494	1652	1710		M	S
18	Benigna von Promnitz	94,979	1648	1702			S
19	Wolfgang Dietrich zu Castell-Castell	94,117	1641	1709			S
20	Heinrich I Reuss zu Kostritz	27,816	1639	1692		M	S
21	Sofia Amalie of Brunswick-Luneburg	11,631	1628	1685			S
22	Johann Friedrich zu Solms-Laubach	35,378	1625	1696		M	S
23	Frederik III of Denmark	11,628	1609	1670		M	S
24	Ludwig Heinrich of Nassau-Dillenburg	92,972	1594	1662			S
25	Katharina zu Sayn-Wittgenstein	92,973	1588	1651			S
26	Maria von Schonburg-Waldenburg	93,904	1565	1628			S
27	Heinrich V Reuss von Plauen	27,829	1549	1604		M	S
28	Anna zu Stolberg	94,448	1548	1599			S
29	Ludwig III zu Lowenstein	34,824	1530	1611		M	S
30	Erika von Waldeck	97,686	1511	1560			S
31	Dietrich V von Manderscheid	97,685	1508	1560			S

- These persons and their ancestors comprise the common ancestors of the Royal Families of Western Europe.

National Communities

How can the analysis be extended to find progenitors for all persons of Western European descent? The success of this research goal depends upon the contents of the Research Genealogy that is being analyzed. The genealogy needs to link Royal Families back to cultural tradition to ensure that all national communities in Europe are represented. The genealogy also needs to include prior kings and the peerages of European countries. Fortunately, information published about the Royal Families provides lineages back to the cultural tradition of each country.

Table 3 lists the national communities present in the Research Genealogy. Each of these national communities is linked to the ancestry of at least one of the Royal Families. All major national communities in Western Europe are included.

As each lineage was entered, a birth year was estimated from the available information if the actual date was not known. This was done to provide a coherent timeline across all the cultural traditions. The only exception was the use of traditional dates defined by James Ussher (Primate of Ireland) for the Davidic King list of Israel. This led to 77 non-biologically feasible dates for age at birth of a child and age at death. The Davidic King list is parsed from the King James Version of the Bible from the books Genesis, Kings, Mathew, and Luke. The traditional explanation is that the names represent tribes instead of individuals. All other dates associated with cultural traditions were constrained to give biologically feasible age ranges for age at birth of children and age at death.

Table 4 lists the time spans for the national communities included in the Research Genealogy, which range from the nineteenth century for the peerage of England back to 1650 bce (before the current era) for the Pharaohs of Egypt. Lineages to these groups were found in Morby's "Dynasties of the World" (Morby, 1989). Lineages for the Pharaohs of Egypt were found in Clayton's "Chronicle of the Pharaohs" (Clayton, 1994).

Lineages for Royal Families can be traced from the present back to 1717 bce, and encompass each successive wave of emigration (Huns, Ostrogoths, Vandals, Lombards, Visigoths, Moors, Mongols, Turks, etc.). Invariably, descendants of each wave of emigration intermarried with the Royal Families. From the present, looking into the past, a unifying ancestry is seen in which ancestors are found in all the cultural groups.

Table 5 summarizes for each given nobility title, the number of persons in the Research Genealogy (#-Persons), the number of ancestors of the Root person (Ancestors), the number of descendants of the ancestral person (Descendants), the number of persons in the Unifying Ancestry (Coreanc), and the number of common ancestors of the Royal Families (Comgroup).

- The Root person for tracing ancestors in Table 5 is President Barack Hussein Obama Jr.
- The Ancestral person from whom descendants are traced is Emperor Charlemagne the Great.
- The Unifying Ancestry is the ancestors of George Alexander Mountbatten-Windsor, known as Prince George of Cambridge.

Table 3 Cultural groups in the Research Genealogy

The Research Genealogy file contains ancestors of persons of European descent
Both historical figures and persons from cultural tradition are included. The following cultural groups are represented:
Anglo Saxons—Kings of Essex, Sussex, Wessex, Mercia, Northumberland, and Kent
Asturia—Imams of Seville
Carolingians—French descendants of Charlemagne
Davidic King list back to Adam and Eve (Ussher chronology)
Eastern Roman Emperors of Byzantium
Egyptian Pharaohs
English peerage to the 1800s
Franks—descendants of Clovis I
German peerage
Irish Monarchs—High Kings of Northern Ui Neill, Southern Ui Neill, Leinster, Meath (Tara), Munster (Cashel), Ulster (Aileach), Breifni, Tyrone (Ulidia, Oriel), Dublin
Latin Kings of Jerusalem during the Crusades
Lombards—Italian and French lineages
Macedonians—Cousins of Alexander the Great
Merovingians—Frankish descendants of Merovech
Ostrogoths—Ancestors of King Theodoric the Great
Persians—Darius the Great
Ptolemaic rulers of Egypt down to Cleopatra
Romans—Western Roman Emperors back to Julius Caesar
Russians—Dukes of Kiev and the Tsars through the Romanovs
Scandinavians—Kings of Norway, Sweden, and Denmark back to Oden
Scots—Kings of Dalriada through Alpin, Kenneth, and MacBeth
Spanish—Kings of Aragon, Castile, Leon, and Navarre
Vandals—Kings Geiseric, Hunneric, and Hilderic
Visigoths—Kings Theodoric, Alaric, and Athanagild
Welsh—Kings of Gwynedd, Deheubarth, and Powys

Table 4 Timeline for cultural groups present in the research genealogy

Monarchs of Ireland	1717 bce–1112
Pharaohs of Egypt	1650 bce–170 bce
Kings of Troy	1540 bce–1130 bce
Kings of Israel	1075 bce–77 bce
Kings of Macedonia	800 bce–212 bce
Kings of Persia	675 bce–628
Goths	362 bce–722
Kings of Franks	76 bce–710
Emperors of Rome	63 bce–718
Kings of Huns	10 bce–692
Kings of Ostrogoths	180–526
Kings of Vandals	268–528
Lombards	275–721
Emperors of Eastern Rome	317–1014
Kings of Visigoths	376–710
Emperors of Byzantium	568–1410
Kings of Sweden	726–Present
Moors—Kings of Asturia	765–854
Kings of Navarre	810–1553
Kings of England	836–Present
Kings of Leon	871–1171
Tsars of Bulgaria	929–1443
Kings of Aragon	1014–1462
Kings of Castile	1016–1425
Kings of Denmark	1043–Present
Latin Kings of Jerusalem	1059–1265
Mongols	1062–1272
Turks—Ottoman Empire	1265–1868
Tsars of Russia	1530–1917
Kings of Netherlands	1772–Present
Kings of Belgium	1790–Present

Table 5 Nobles in the Research Genealogy

Type	Title	#-Persons	Ancestors	Descendants	Coreanc	Comgroup
24	Emperor	231	90	94	100	100
23	Monarch	178	100	0	100	92
22	Pharaoh	50	34	0	34	34
21	U.S. President	46	0	46	0	0
20	Tsar	62	2	26	11	5
19	Saint	127	54	17	54	54
18	Pope	79	0	31	1	1
17	King	2375	934	557	1051	987
16	Queen	191	81	129	109	99
15	Prince	2924	182	2052	326	288
14	Princess	1919	146	1645	282	230
13	Duke	2758	332	2201	696	559
12	Duchess	714	53	653	192	135
11	Khalif	70	0	1	0	0
10	Earl	2092	317	1801	615	223
9	Count	8828	1583	5061	3235	3051
8	Countess	3673	268	2333	947	811
7	Marquess	889	40	494	96	56
5	Marquessa	142	3	98	39	29
4	Baron	3982	202	1670	554	268
3	Baroness	925	20	279	89	41
2	Lord	4999	850	3188	1929	812
1	Knight	4497	632	1902	1588	265

Notable information about the Research Genealogy from Table 5 includes:

- The Research Genealogy that is analyzed contains 330,610 persons.
- The total number of titled persons is 41,751 and contains the peerage of England down to the 1800s.
- There are 2375 Kings in the Research Genealogy.
- The descendants of Charlemagne include 557 Kings.
- All the U.S. Presidents are included in the Research Genealogy.

- The ancestors of Barack Obama include 934 Kings, and the ancestors of Prince George of Cambridge include 1051 Kings. A person whose ancestry connects to the Unifying Ancestry will expect to have descents from hundreds of Kings.
- Except for nobles associated with the English peerage, the number of nobles in the common ancestors of the Royal Families is only slightly smaller than the number of nobles in the Unifying Ancestry. The Unifying Ancestry is strongly tied to the common ancestors of the Royal Families, despite being the ancestors of Prince George of Cambridge.

Other groups that are included in the Research Genealogy are:

- Signatories of the Magna Charta
- Spouses of the U.S. Presidents
- U.S. Vice Presidents
- Signers of the U.S. Declaration of Independence
- Signers of the U.S. Constitution
- A total of 1986 additional historically notable persons such as politicians, inventors, businessmen, clergy, professors, athletes, actors, singers, etc.

References

Bridges, E. (1970). *Collins's Peerage of England*. AMS Press Inc

Clayton, P. A. (1994). *Chronicle of the Pharaohs*. Thames and Hudson.

Moore, R. (2020). *Core genealogy: A genealogical history of the modern world*. Chapel Hill. www. coregen.center.

Morby, J. E. (1989). *Dynasties of the world*. Oxford University Press.

Palmer, R. R., & Colton, J. (1952). *A history of the modern world*. Alfred A Knopf.

Schwinnicke, D. (1995). *Europaische Stammtafeln: Stammtafeln zur Geschichte der Europaischen Staaten, Neue Folge*. J. A. Stargardt.

Wentworth, J. (1878). *Wentworth genealogy*. Little, Brown, and Company.

Williamson, D. (1991). *Debrett's king and queens of Europe*. Salem House Publishers.

Research Genealogy Knowledge Base

The Research Genealogy is published as a Gedcom version 5.5 formatted file. The Core-Gen3 genealogy analysis workbench parses a Gedcom file to build a graph database that organizes the information extracted from the communications from the past (i.e., sources). The database tracks sources, familial relationships between persons, and provides procedures for evaluating metrics for Consistency, Correctness, Closure, Connectivity, Completeness, and Coherence. Each person is annotated with a list of the sources that provided information about the person. Familial relationships are analyzed to derive a Unifying Ancestry. The Unifying Ancestry is analyzed to find Progenitors for persons of Western European descent. The Progenitors are then linked to an external knowledge base consisting of the historically notable persons defined in Palmer's History of the Modern World (Palmer & Colton, 1952).

Research Genealogy—Consistency

Information from 1738 sources was used to construct the Research Genealogy. The sources initially included family bibles, family histories, census data, history books, and published genealogies. Over time, the type of sources expanded to include web sites, cultural traditions, and oral traditions such as the Heimskringla and Orkneyinga sagas (Sturluson, 1964). The goal is to construct a knowledge base that serves as an index (similar to the index of an archive) to the information content within all the referenced sources.

A major question is whether a trustworthy genealogy can be generated from a collection of arbitrary sources. This is very similar to the challenge of interpreting information from multiple communications from the past. Each communication may contain both accurate and inaccurate information. By analyzing relationships between information

R. W. Moore, *Trustworthy Communications and Complete Genealogies*, Synthesis Lectures on Information Concepts, Retrieval, and Services,
https://doi.org/10.1007/978-3-031-16836-9_3

elements extracted from each communication, a knowledge base can be assembled. The knowledge base properties can then be evaluated to decide whether the resulting genealogy analyses can be trusted.

Techniques that were used include validating the internal consistency of each source, comparing information between multiple sources, and applying correctness evaluation metrics. Discrepancies were annotated for each source. The attributes for each person were set to the information retrieved from the most trustworthy sources which typically provided the most detailed information.

Trustworthy sources were found for the following types of communication:

- Family ancestries. An example is *"Ancestry of Charles II, King of England, A Medieval Heritage"* by Hansen and Thompson (2012).
- Descendants of family progenitors. An example is *"The Book of Kings: A Royal Genealogy"* by Arnold McNaughton (1973).
- Compiled genealogies. An example is *"Europaische Stammtafeln: Stammtafeln zur Geschichte der Europaischen Staaten, Neue Folge"* by Schwinnicke (1995).
- Genealogy lineages. An example is *"The Magna Charta Sureties"* by Weis (1982).
- History books. An example is *"A History of the Crusades"* by Runciman (1962).
- Web sites. An example is "FindaGrave". http://www.findagrave.com.

By including both trustworthy and untrustworthy sources, the index can identify when the untrustworthy sources differ from the trustworthy sources.

A major concern was the construction of normalized name spaces for person names, dates, titles, and locations across all the sources. Alternate names were recorded as aliases, and dates were preferentially listed using the Julian calendar. Place locations were preferentially listed using the current country designation. Table 1 provides a distribution by century for the number of persons referenced by the ten sources with the largest number of referenced persons. The column labeled Cent identifies the starting year of the century. The column labeled Total lists the total number of persons in the Research Genealogy that were born in the century. Note that persons born before the current era are included in the number for the first century. The ID of the source is listed in the row labeled "Cent". Thus, for source 224S (6th column from the left), a total of 4 persons were referenced in the ninth century, 41 persons were referenced in the tenth century, and 255 persons were referenced in the eleventh century. The row labeled Ave. lists the average birth year for all persons referenced by the source. The row labeled Tot. lists the total number of persons referenced by the source that are included in the Research Genealogy.

Notable information in Table 1 includes:

- The number of persons born before 0 AD is 2010. These persons predominantly represent cultural traditions. Exceptions include the Pharaohs of Egypt (documented by

Table 1 Number of persons referenced in each century for the 10 sources with the most referenced persons

Cent	711S	589S	1376S	260S	224S	258S	1146S	216S	150S	956S	Total
0	129	0	0	0	0	1	0	0	0	4	2442
100	46	0	0	0	0	0	0	0	0	6	384
200	47	0	0	0	0	0	0	0	0	5	389
300	65	0	0	0	0	0	0	0	0	15	527
400	55	0	0	0	0	0	0	2	22	6	824
500	68	0	0	0	0	0	1	35	94	16	912
600	65	0	0	0	0	0	0	66	68	20	936
700	74	10	0	0	0	5	0	119	86	18	1117
800	93	8	0	0	4	8	8	525	197	27	2021
900	182	0	0	0	41	34	15	834	311	77	3534
1000	652	0	45	23	255	100	61	1402	448	181	6666
1100	1530	0	55	398	785	257	121	1767	525	333	11,021
1200	2508	7	165	1363	1987	493	334	1835	743	494	16,443
1300	3207	268	549	3021	1788	981	456	1793	918	541	19,940
1400	4790	1529	1692	5497	1430	1986	595	1071	836	770	27,148
1500	6489	4319	3015	5126	1909	3160	1284	424	1360	1417	41,346
1600	9093	8215	8592	2428	1743	3130	3683	350	1802	2173	66,507
1700	9189	12,613	7411	53	1740	1540	3178	458	1169	2125	65,985
1800	7384	3346	2921	0	2247	858	2059	671	1027	1417	41,259
1900	1478	0	465	0	767	263	728	263	626	265	21,155
2000	0	0	1	0	0	0	0	0	0	0	54
Ave	1588	1684	1671	1465	1545	1568	1673	1287	1497	1595	1571
Tot	47,144	30,315	24,911	17,909	14,696	12,816	12,523	11,615	10,232	9910	330,610

Cent is the start of the century, followed by source numbers
Ave. is the birth year averaged across all referenced persons
Tot. is the total number summed across all referenced persons
There are 330,610 persons in the genealogy

hieroglyphs on the walls of the tombs in the Valley of the Kings), and the Emperors of Rome.

- The number of persons born before 1000 AD is 13,087. Most of these persons are defined by cultural tradition, but they do include historical persons such as Charlemagne, Clovis I, and Emperors of Rome.
- Persons born after 1200 AD are almost entirely parsed from an historical context.
- By tracking the number of persons born each century that are referenced by a source, a decision can be made for whether the source focuses on cultural tradition or historical events. Thus, source 260S mainly references historical events, while source 150S includes cultural tradition.

An effort was made to find information from multiple sources about each person. All person names and all dates were annotated with the reference number of the source, shown in the row labeled "Cent". The sources were numbered in the order in which they were read. Since the construction of the Research Genealogy started in 1990, the earlier source numbers reference books. As the World Wide Web matured, additional sources were used from genealogy web sites. All information was manually added to the Research Genealogy using the Reunion genealogy program.

Table 2 lists the reference for each source identified in the top row in Table 1. Once the progenitors for the Royal Families were identified, additional sources were used to fill in gaps and verify familial relationships.

A total of 515,171 source references are made in the Research Genealogy, averaging 1.55 sources per person. The largest number of sources for a person was 41 for King Henry I Beauclerc. For sources that comprised multiple book volumes, a separate source number was created for each volume. Among the top 10 sources are three volumes of Europaische Stammtafeln; source ID 150 is Volume 1.1, source ID 216 is Volume 2, and source ID 224 is Volume 3.1. The total number of references to 27 volumes of Europaische Stammtafeln is 79,796 persons. The total number of references to 9 volumes

Table 2 Top 10 sources for the construction of the Research Genealogy

1-711S	Web Site, Geni, Geni. http://www.geni.com/. 10 Aug 2012
2-589S	Web Site, Harrison, Bruce, The Family Forest Descendants of King Edward III of England and Queen Philippa of Hainault. http://www.familyforest.com. Change Date 8 JAN 2018
3-1376S	Web Site, FamilySearch, FamilySearch. https://ancestors.familysearch.org. Change Date 26 NOV 2019
4-260S	Book, Richardson, Douglas, Plantagenet Ancestry, Everingham, Kimball G., Genealogical Publishing Company, 2004
5-224S	Book, Isenburg, Wilhelm, Europaische Stammtafeln, Verlag von J. A. Stargardt, 1956, Vol 3.1, CS 616 I7 Change Date 17 OCT 2020
6-258S	Book, Roberts, Gary Boyd, The Royal Descents of 600 Immigrants to the American Colonies or the United States, Genealogical Publishing Company, Incorporated
7-1146S	Web Site, Dowling, Tim, Tim Dowling's Family Tree. http://gw.geneanet.org/tdowling. Change Date 9 NOV 2016
8-216S	Book, Schwennicke, Detlev, Europaische Stammtafeln, Stammtafeln zur Geschichte der Europaischen Staaten, Verlag von J. A. Stargardt, Vol 2, Change Date 17 OCT 2020
9-150S	Book, Schwennicke, Detlev, Europaische Stammtafeln, Neue Folge, Vittorio Klostermann, Frankfurt am Main, 1998, Vol I.1, Change Date 17 OCT 2020
10–956	Web Site, Wikitree, Wikitree. http://www.wikitree.com, 1/24/2016

Table 3 The 20 countries with the most persons referenced in the Research Genealogy

Top Countries	Number of persons
United States	51,098
England	33,745
Germany	14,821
France	9244
Scotland	4176
Netherlands	2849
Italy	1935
Ireland	1917
Belgium	1383
Spain	1226
Denmark	1212
Austria	1193
Wales	1126
Sweden	1072
Canada	966
Poland	896
Russia	743
Czech Republic	503
Switzerland	498
Norway	441

of Collins's Peerage of England is 53,088 persons. These are the two major sources used to assemble the Research Genealogy.

The construction of a normalized name space for birth locations required the review of every birth location, the addition of a country name if missing, and the tracking of changes as national boundaries shifted. Table 3 lists the number of persons present in the Research Genealogy for the top 20 countries for which a birth location has been entered. The distribution of persons by country reflects the addition of notable United States residents to the Research Genealogy (politicians, businessmen, inventors, actors, and professors), the inclusion of the persons from the Peerage of England, and the inclusion of persons from the Noble houses of Germany. Currently, 135,355 persons in the Research Genealogy have defined birth locations representing 137 countries.

For each person in the Research Genealogy, the following information was parsed from the Gedcom file:

- Last name.
- First name.
- Birth year.
- Person ID.
- Familial relationship to another person (spouse family ID, parent family ID, child family ID).
- Source identifier for the information.
- Alias if known by another name.
- If known, birth date and birth location.
- If known, birth country location.
- If known; baptism date and location, marriage date and location, death date and location, and burial date and location.
- If known; title, residence location, education, alias, occupation, and cause of death.
- If known, prefix and suffix.
- If available, notes defining events during the person's life.
- If death and birth years are known, the age is calculated.

Consistency was verified by sorting each variable and verifying that no required attributes were missing.

Research Genealogy—Correctness

Unfortunately, every source contained errors or discrepancies in names, dates, locations, titles, and familial relationships. Since information tends to appear multiple times in a source (as a child, as a spouse, and as a parent), sources were rated as more reliable if the same information appeared each time for a person. Discrepancies between sources were annotated, and the source providing more detailed information was usually taken as more reliable. Comparisons were made between genealogies that compiled lineages to notable persons, genealogies that compiled descendants of a notable person, genealogies that compiled ancestors of a notable person, and genealogies that compiled descendants of a family progenitor.

Not even primary records were accurate. My mother was born in Saskatchewan, Canada. She and her cousin disputed whether her birth certificate listed the date of her birth or the date the birth was recorded at the post office. Common errors encountered across the sources included typographical errors in names, transcription errors between multiple entries of the information, listing the Baptism Date as the birth date, listing the Marriage Date as the birth date, assigning the birth of a child to the wrong spouse or to a sibling, missing generations, alternate numbering for titles, alternate birth locations, alternate spelling of the person's name, etc. Since sources were used from multiple countries, the spelling of the name varied depending upon the language being used.

In addition to comparing information across sources, a method that was source independent was needed to check every entry. A biological consistency check was made to ensure that the age at marriage, age at birth of a child, and age at death fell within biologically feasible ranges. This check was applied across both historical records and cultural traditions. The only persons that failed the age range checks came from the use of traditional dates from the Davidic King list, which assigns a birth year of 4004 bce to Adam and Eve. Otherwise, all ages fell within biologically feasible ranges. Table 4 provides a distribution by century for the average age at birth of a child (male-gen, female-gen), average age at marriage (male-mar, female-mar), and average age at death (male-age, female-age) for men and women born within the century. Only dates in the current era were analyzed.

The male and female generation averages were used to estimate missing birth years. Table 4 indicates that on average:

- The average number of years per generation has slowly increased from 29 to 34 for men, and from 24 to 29 for women. The average is calculated across all children born in the century. When estimating birth years, an average value of 30 is used for the number of years per generation.
- On average, men marry women that are 2–6 years younger. This effect is increased by the second and third marriages of an individual.
- The average age at death is 56 for men and 57 for women. The average strongly depends upon the number of children that are identified that die at an early age.
- The average age at marriage includes marriage contracts that were created for children as young as one year old in the Middle Ages.

Table 5 provides a distribution by 5-year age interval of the number of persons who had children (male-gen, female-gen), were married (male-mar, female-mar), or died (male-age, female-age) within the age interval for both men and women.

The age ranges in the Research Genealogy are biologically feasible for persons born in the current era:

- The Research Genealogy contains 288 persons who lived more than 100 years. Note that the current oldest living person is 118.
- The peak in the age at death occurs between 60 and 65 for men and between 65 and 70 for women.
- The peak in the age at marriage is between 20 and 25 for men and between 15 and 20 for women, but more men marry at older ages.

Table 4 Distribution by century of the average age at birth of a child, marriage, and death for men and women

Year	Male-gen	Female-gen	Male-age	Female-age	Male-mar	Female-mar
0	30	25	71	50	36	24
100	29	23	54	47	19	19
200	28	24	57	50	28	26
300	30	25	54	48	32	24
400	30	26	53	52	31	28
500	30	26	51	46	29	24
600	30	25	52	52	25	20
700	29	26	53	51	29	20
800	30	25	50	51	29	22
900	30	25	51	52	28	22
1000	31	26	51	50	29	22
1100	31	26	51	52	29	22
1200	31	26	50	50	28	21
1300	31	26	51	50	28	22
1400	32	27	54	53	29	23
1500	32	27	54	53	29	24
1600	33	28	56	55	29	23
1700	33	29	59	58	28	23
1800	34	28	62	63	30	24
1900	32	27	59	66	29	25
2000	35	33	0	0	0	0
Average	32	28	56	57	29	23

Gen average age of parent at birth of all children within century
Age average age at death for all persons born within century
Mar average age of person at marriage for marriages in century

- There are 383 women who had children in their 50s and 1571 men who had children after 59. Note that the oldest recorded age for a woman at birth of a child is 66.

Research Genealogy—Closure

There may not be sufficient information available about a person's ancestry to find a cousin relationship to the unifying ancestry. Instead, connections may require the traversal

Table 5 Distribution of number of persons who had children, or died, or married per 5-year age interval

Age	Male-gen	Female-gen	Male-age	Female-age	Male-mar	Female-mar
0	0	0	3487	2503	8	15
5	0	0	696	445	52	114
10	167	524	579	374	470	1600
15	4975	14,779	1035	679	3025	11,199
20	24,394	51,591	2044	1611	12,501	16,322
25	57,222	57,229	2942	2608	11,595	7877
30	54,434	40,697	4124	3356	6772	3570
35	37,025	25,107	5117	4123	3816	1696
40	24,513	9912	6304	4607	2374	815
45	13,519	2218	7487	4530	1484	501
50	6411	353	8327	4844	955	282
55	2744	30	9255	5330	558	178
60	1122	0	9827	5785	343	97
65	348	0	9641	6053	206	39
70	87	0	9350	6283	118	28
75	12	0	8253	5945	43	5
80	2	0	6298	5068	12	1
85	0	0	3942	3545	6	0
90	0	0	1876	1945	1	0
95	0	0	689	902	0	0
100	0	0	113	149	0	0
105	0	0	5	16	0	0
110	0	0	3	2	0	0
115	0	0	0	0	0	0
Total	226,975	202,440	101,425	70,705	44,339	44,339

of multiple extended families. Each extended family consists of the ancestors, descendants, and relatives of a "root" person. For instance, two persons may be connected through a cousin of a spouse of a cousin. Closure requires that a connection can be found between any two persons in the genealogy. There should be no disconnected islands.

To verify that the Research Genealogy does not contain any islands of disconnected persons, a partitioning algorithm was written that partitions the genealogy into extended families. The algorithm uses the following steps:

- Find all the ancestors, descendants, and relatives for the root person.
- Generate a list of all unrelated spouses of the descendants and relatives.
- For each unrelated spouse identify the members of their extended family, excluding persons already in an extended family.
- When the list has been processed, this generates a ring of extended families around the original partition.
- Generate a new list of all unrelated spouses from the last ring of extended families and iterate.
- This produces successive rings of extended families about the original partition.
- At the end of the process, all persons in the genealogy should be a member of an extended family. Persons without a partition number are disconnected from the rest of the genealogy.

Table 6 lists the partition numbers for each ring of extended families. The analysis uses Prince George of Cambridge as the root person for the first extended family. Each ring of extended families corresponds to another degree of separation. The column labeled Degree denotes the number of degrees of separation. "0" references the initial extended family for Prince George. "1" denotes the first ring of extended families around the initial extended family. The Partition-range defines the partition numbers associated with each extended family in the ring of extended families. #Persons lists the total number of persons in the ring. %Total lists the cumulative percentage of the number of persons in the Research Genealogy found across all interior rings.

Notable information in Table 6 is:

- The extended family for Prince George of Cambridge has 187,609 persons, over half of the Research Genealogy.
- There are 58,413 families in the first ring (one degree of separation), representing 58,413 unrelated spouses of the cousins and descendants of Prince George of Cambridge.
- 97.61% of the persons in the Research Genealogy are within two degrees of separation.
- The lineage for the most distantly connected person in the Research Genealogy traverses 11 additional extended families to connect to Prince George.
- Also, the analysis identifies 54,813 unrelated spouses for which no information about parents is available. 38% of the families in rings 1–11 have a single person in their extended family.
- A total of 71,179 extended families were identified. The total number changes depending upon who is selected as root of the first extended family.
- All persons in the Research Genealogy are members of an extended family that can be linked to the relatives of Prince George of Cambridge through multiple degrees of separation.

Table 6 Partition ranges for each degree of separation (ring of families)

Degree	Partition range	#Persons	Total (%)
0	1–1	187,609	56.75
1	2–58,414	117,852	92.39
2	58,415–68,444	17,236	97.61
3	68,445–69,786	3755	98.74
4	69,787–70,489	2397	99.47
5	70,490–70,877	1079	99.79
6	70,878–71,059	375	99.91
7	71,060–71,107	131	99.95
8	71,108–71,141	100	99.98
9	71,142–71,160	48	99.99
10	71,161–71,176	25	100
11	71,177–71,179	3	100

Each degree of separation corresponds to another ring of families
Partition-range lists the partition numbers in the ring of families
#Persons is the total number of persons in the ring of families
%Total is the cumulative percentage of persons across all inner rings

Research Genealogy—Connectivity

Joseph Chang, a Yale statistician, has calculated how far back in time ancestry must be traced to find a person who is the ancestor of all persons alive today (Chang, 1999). His analysis is based on population genetics using a two-parent model, assuming the size of the population remains constant. Chang found that the number of generations back to the common ancestor is

$$N_g = Log2(population\text{-}size) \qquad (1)$$

Here $Log2(x)$ is the logarithm to the base 2, which counts the number of factors of two present in the function argument x. For a population of 200 million persons, a common ancestor should be found in less than 28 generations ($2^{28} \sim 256$ million). An estimate for the number of generations in the past when all persons are common ancestors of everyone alive today is

$$N_c = 1.77 \, Log2(population\text{-}size) \qquad (2)$$

Thus 49 generations in the past, any person that was alive is a common ancestor of all members of the current Western European population of 200 million persons. These estimates depend upon multiple factors:

- Since historically the population is not constant but a factor of 10 smaller in the past than the current population, this leads to an overestimate of the number of generations.
- The analysis assumes intermarriage across all members of the population and does not consider marriage restricted within sub-groups. This leads to an underestimate of the number of generations if the intermarriage period between sub-groups is greater than a generation.
- The analysis assumes perfect knowledge of the lineages connecting the population to the progenitors.

If we take 49 generations as the upper bound for the appearance of a common ancestor, it is possible to find such a person using lineages for the Royal Families of Europe. In the Research Genealogy there are 99 persons who are the 47th Great-GrandParents of Prince George of Cambridge. An example is:

- Helen of the Cross, born about 248 with 2.6 trillion ascents from Prince George.

Helen is a candidate for a progenitor for all persons of Western European descent. The challenge then becomes whether sufficient information is available in the communications from the past to validate this analysis. Identifying progenitors requires tracking lineage coalescence back at least 49 generations. Finding the progenitors will need to be augmented with finding all the supporting lineages. The demonstration can be simplified through the construction of a Unifying Ancestry for Western Europeans.

Multiple graph traversal algorithms were developed to track the connectivity that is needed for creating a Unifying Ancestry. The Unifying Ancestry should include:

- An exponentially growing number of ascents as lineages are traced back to a progenitor.
- A small number of generations from the root person of the genealogy to the first ancestor with multiple ascents.
- A set of unrelated progenitors from whom Western Europeans alive today can trace their ancestry.

Finding a lineage to one member of the Unifying Ancestry should lead to familial relationships to the other members of the Unifying Ancestry. This requires a high degree of connectivity between the members of the Unifying Ancestry which leads to a requirement for a high degree of lineage coalescence. Within the CoreGen3 workbench, there are four ways to demonstrate the exponentially growing lineage coalescence within the Research Genealogy.

1. Descendant fractional distribution
2. Global connectivity metric

3. Own cousin relationship
4. Local connectivity metric.

Connectivity—Descendant Fractional Distribution

An implication of lineage coalescence is that a significant fraction of the Western Europeans alive today should be able to trace their ancestry to Charlemagne. An interesting question is whether the ancestry is non-negligible, with multiple genes inherited from Charlemagne. Since there are about 20,000 genes, this requires the fraction of the ancestry from Charlemagne to be greater than 0.05%.

The descendant fractional ancestry is calculated by:

- Looping over the descendants of a family
- For each child, setting the descendant fractional distribution to half of the sum of the descendant fractional distributions of the parents.

The result of the analysis is shown in Table 7 for the descendants of Charlemagne. For each descendant generation, the person with the maximum descendant fractional distribution is identified (column DNA). This is compared with the minimum descendant fractional distribution which is equal to $1/(2^{Gen})$ where Gen is the number of generations back to Charlemagne (column minDNA). The column labeled Gen counts the number of generations from Charlemagne to the descendant. The column labeled Birth lists the birth year.

The high degree of connectivity is shown by the fact there are persons born in the twentieth century whose descendant fractional distribution is greater than 0.1%. This is more than 10 million times larger than the minimum value. The number of descendants of Charlemagne contained in the Research Genealogy is 150,492 persons. Of these, 19,313 are among the ancestors of Prince George of Cambridge. Note that 45% of the persons in the Research Genealogy are descendants of Charlemagne.

King Felipe VI of Spain traces 0.203% of his ancestry back to Charlemagne. Thus, he may have inherited genes from Charlemagne. Note that 9 of the 33 persons are Kings, indicating a very high rate of intermarriage between Royal Families. The current Kings and Queens of Western Europe are all descendants of Charlemagne. In the Research Genealogy, 40% of the Kings born after 750 AD are his descendants, as are 61% of the Emperors born after 750 AD.

Table 7 Maximum descendant fractional distribution found per generation versus the minimal expected descendant fractional distribution

Gen	minDNA	DNA	Birth	Person
1	5.00E−01	5.00E−01	773	King Pippin III of Italy
2	2.50E−01	2.50E−01	797	King Bernard de Vermandois
3	1.25E−01	2.50E−01	836	Gosselin III du Maine
4	6.25E−02	1.25E−01	850	Witbert of Nantes
5	3.12E−02	1.02E−01	918	Geraud I de Limoges
6	1.56E−02	7.03E−02	960	Almodis de Limoges
7	7.81E−03	5.86E−02	960	Adele de Senlis
8	3.91E−03	4.00E−02	1004	Adele de Coucy
9	1.95E−03	3.42E−02	1057	Agnes de la Fere
10	9.77E−04	2.57E−02	1070	Gertrude of Flanders
11	4.88E−04	2.36E−02	1099	Rosamunda of Flanders
12	2.44E−04	1.61E−02	1170	Isabelle of Hainault
13	1.22E−04	1.43E−02	1187	King Louis VIII of France
14	6.10E−05	1.18E−02	1215	King Louis IX of France
15	3.05E−05	1.07E−02	1248	Blanche d'Artois
16	1.53E−05	8.73E−03	1303	Catharine II of Valois
17	7.63E−06	8.23E−03	1312	Joan II of France
18	3.81E−06	7.81E−03	1332	King Charles II d'Evreux
19	1.91E−06	5.81E−03	1410	Philippe of Burgundy
20	9.54E−07	5.54E−03	1432	Royan de Bretagne
21	4.77E−07	4.99E−03	1496	Claude I de Lorraine
22	2.38E−07	4.27E−03	1553	King Henry III of Navarre
23	1.19E−07	3.68E−03	1587	Victor Amadeus I of Savoy
24	5.96E−08	3.32E−03	1636	Adelaide Henrietta Maria of Savoy
25	2.98E−08	3.28E−03	1660	Maria Anne-Christine Victoria de Bavaria
26	1.49E−08	3.26E−03	1692	Joseph Ferdinand Leopold of Bavaria
27	7.45E−09	2.95E−03	1710	King Louis XV of France
28	3.73E−09	2.63E−03	1798	Marie Caroline Ferdinanda Louise of Bourbon-Two Sicilies
29	1.86E−09	2.57E−03	1839	Karl Salvator of Habsburg-Tuscan
30	9.31E−10	2.47E−03	1862	Pedro de Alcantara de Bourbon y Bourbon
31	4.66E−10	2.41E−03	1903	Helena of Habsburg-Lorraine

(continued)

Table 7 (continued)

Gen	minDNA	DNA	Birth	Person
32	2.33E−10	2.12E−03	1938	King Juan Carlos I Alfonso Vittorio Maria of Spain
33	1.16E−10	2.03E−03	1968	King Felipe VI of Spain

Gen generation number
minDNA minimum descendant fractional distribution = $1/(2**GEN)$
DNA descendant fractional distribution
Birth birth year
Person person with the largest descendant fractional distribution in the generation

Connectivity—Global Connectivity Metric

A way is needed to identify additional potential progenitors. They should be unrelated to Charlemagne and have a large global connectivity. A direct measure of connectivity is to count the number of ascents from the root person to an ancestral person. Each ascent represents a unique path through the genealogy. The larger the number of ascents, the more paths that exist through the genealogy to the ancestral person.

The global connectivity metric identifies the persons in the genealogy with the largest number of ascents over the smallest number of generations. A metric which calculates this is:

- Global connectivity metric = Log2(number of ascents)/(number of generations)

The values of the global connectivity metric greater than 0.975 for the ancestors of Prince George of Cambridge are shown in Table 8. The column labeled #Gen lists the number of generations from Prince George of Cambridge to the ancestor. The column labeled #Ascents lists the number of unique paths through the genealogy connecting Prince George and the ancestor. The column labeled Metric contains the global connectivity metric. The column labeled Relation contains the relationship of the person to Charlemagne.

The relation 12G-GFather means 12th Great-Grandfather, and 12G-GMother means 12th Great-Grandmother. A global metric value of 1 means the number of ascents doubles on average each generation. This is possible when there are two or more children each generation who are members of the common ancestors. The number of ascents is measured in the tens of billions to tens of trillions for the persons with the highest global connectivity metric. Note that $2**44$ is 17.5 trillion. The birth years range from 255 to 688 AD and the number of generations ranges from 34 to 47.

The person with the largest number of ascents from Prince George of Cambridge in the Research Genealogy is Tao I of Egypt with 245 quadrillion ascents. A quadrillion is one thousand trillion, or one million billion. The number for Tao I is so large because the

Table 8 Analysis of the global connectivity metric for the ancestors of Prince George of Cambridge

	Person	Birth	#Gen	#Ascents	Metric	Relation
1	Theuderic III of Neustria	653	35	3.36E+10	0.999	G-GFather
2	Clotilda of Brabant	652	35	3.36E+10	0.999	G-GMother
3	Marcomir I of Franks	360	43	7.20E+12	0.9933	9G-GFather
4	Hildegonde de Cologne	375	43	7.20E+12	0.9933	9G-GMother
5	Blesinde von Alemannien	279	46	5.47E+13	0.9921	12G-GMother
6	Begga of Brabant	613	36	5.48E+10	0.991	2G-GMother
7	Ragaises von Toxandrien	270	46	5.25E+13	0.9908	12G-GFather
8	Ansegisel of Brabant	602	36	5.40E+10	0.9904	2G-GFather
9	Clodius I of Franks	340	44	1.30E+13	0.9901	10G-GFather
10	Blessinde of Franks	340	44	1.30E+13	0.9901	10G-GMother
11	Bertrada I of Neustria	680	34	1.31E+10	0.9886	GMother
12	Amalaberge of Ostrogoths	386	43	6.07E+12	0.9876	9G-GMother
13	Blesinde of Franks	370	43	5.80E+12	0.986	9G-GMother
14	Widelphe of Saxony	380	43	5.80E+12	0.986	9G-GFather
15	Charles Martel	688	34	1.19E+10	0.9846	GFather
16	Asilia of Lombards	308	45	2.02E+13	0.9822	11G-GMother
17	Weldelphe of Saxony	408	42	2.57E+12	0.9816	8G-GMother
18	Bisinus II of Thuringia	445	41	1.23E+12	0.9795	7G-GFather
19	Menia of Cologne	445	41	1.23E+12	0.9795	7G-GMother
20	Chrocus I von Alemannien	255	47	6.96E+13	0.9784	13G-GFather
21	(daughter) of the Alamans	255	47	6.96E+13	0.9784	13G-GMother
22	Chrotlind of the Upper Rhine	675	35	2.02E+10	0.9781	G-GMother

Metric = log2(# number of ascents from core root person)/(# of generations from core root person)
#Gen number of generations from root of core genealogy to person
#Ascents number of ascents from root of core genealogy to person
Relation familial relationship to Charlemagne

Pharaohs of Egypt married half-siblings, doubling the number of ascents each generation. However, this is 148 generations in the past giving a global connectivity metric of 0.39. Charlemagne has 3.24 billion ascents but over only 33 generations, resulting in a global connectivity metric of 0.957.

The global connectivity metric demonstrates why a descent from Charlemagne is sought as a measure of completeness. The persons with higher connectivity values listed in Table 8 are all ancestors of Charlemagne. An ascent to Charlemagne can be extended to an ascent to all the persons in Table 8.

Charlemagne is a candidate for a progenitor for persons of Western European descent. However, the number of generations along an ascent to Charlemagne is as small as 33, much less than the analytic estimate of 49 generations to a progenitor. An alternate way is needed to calculate connectivity for identifying progenitors.

Connectivity—Own Cousin Relationship

A third method for calculating connectivity within a genealogy is to evaluate the own cousin relationship of each ancestor. You are your own cousin. You have an ancestor to which you have two ascents. You can calculate your own cousin relationship by tracing your ancestry up the first ascent to the ancestor, then calculating your cousin relationship coming back down the second ascent. Figure 1 shows this calculation for a person who is his own 5th cousin. The own-cousin lineage coalescence point is the closest ancestor to which you have two ascents.

The own-cousin relationship defines the minimum number of generations that ancestry must be traced to find the first lineage coalescence.

Table 9 shows a distribution of the common ancestors of the Royal Families of Western Europe for each century with a specified own-cousin relationship. The column labeled Year is the starting year of the century. The column labeled #OwnCos lists the total number of persons with own-cousin relationships born in the century. The columns labeled 1C–10C contain the number of persons with that cousin relationship degree. The row labeled Total is the total number of common ancestors with the given cousin relationship.

To interpret the table, note that in the row labeled Total, there are 40 common ancestors that are their own 1st cousin and 409 common ancestors that are their own 2nd cousin. A total of 3438 ancestors are their own 5th cousin.

After 1000 AD, the peak in the distribution of own-cousin relationships occurs around 5th cousin. In earlier centuries, the peak does shift to the 4th and then the 3rd cousin.

Fig. 1 Own cousin relationship

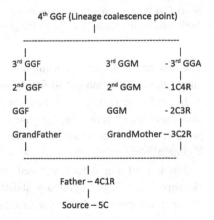

Table 9 Distribution of own cousin relationship by century for common ancestors of royal families of Western Europe

Year	#Owncos	1C	2C	3C	4C	5C	6C	7C	8C	9C	10C
0	436	14	52	45	41	28	18	18	15	13	11
100	141	0	9	23	17	15	11	6	5	5	5
200	130	1	5	3	10	19	20	12	10	7	4
300	207	2	16	35	19	8	7	18	16	20	13
400	310	4	28	46	58	27	18	9	6	11	25
500	339	0	16	58	82	61	30	22	5	3	1
600	335	2	28	58	59	62	35	21	17	10	5
700	388	0	37	90	78	44	40	21	12	12	12
800	632	4	44	120	162	136	60	24	18	9	7
900	1013	5	39	135	262	213	143	78	49	32	23
1000	1775	5	37	147	325	414	332	223	106	57	43
1100	2317	1	37	150	313	515	462	312	206	127	62
1200	2854	0	14	121	468	780	555	355	188	127	83
1300	2072	2	12	96	417	532	390	200	147	94	60
1400	1263	0	8	78	262	326	209	123	87	54	32
1500	698	0	5	61	212	201	87	42	29	17	12
1600	270	0	18	80	97	56	11	2	3	2	1
1700	38	0	4	16	16	1	1	0	0	0	0
1800	0	0	0	0	0	0	0	0	0	0	0
1900	0	0	0	0	0	0	0	0	0	0	0
2000	0	0	0	0	0	0	0	0	0	0	0
Total	15,218	40	409	1362	2898	3438	2429	1486	919	600	399

Year is the start of the century. Century 0 includes all persons born before 0 AD
#OwnCos is the number of common ancestors who have an Own Cousin relationship
Additional columns are the # with the given own cousin relationship

The median value is 6th cousin, and the average value is 7th cousin since there is a long tail of the distribution out to 66th own-cousin relationship driven by lineages to the stem of Ireland. The number of generations to the own-cousin lineage coalescence point is one plus the cousin relationship. The peak in number of generations to the lineage coalescence point occurs at 6 generations, the median value at 7 generations, and the average value at 8 generations.

For a Unifying Ancestry based on the common ancestors of the Royal Families of Western Europe, the summary statistics are given below. Note that a Core child is a common ancestor of the Royal Families.

- Total number of common ancestors of Royal Families   28,124
- Total number with Own Cousin relationship   15,218
- Total number with no parents   6425
- Total number with > 1 Core child   6700
- Total number of closest lineage coalescence points   2119
- Average own cousin relationship   6.449
- Average number of generations to lineage coalescence point   7.449
- Average number of lineages combined at each lineage coalescence point   16.033
- Average number of core children for lineage coalescence points   2.562
- Average number of core children when 2 or more core children   2.814

For a Unifying Ancestry based on the common ancestors of the Royal Families, half of the members have an own-cousin relationship. About a quarter of the Unifying Ancestry members have no known parents. The closest own-cousin relationships terminate on only 2119 lineage coalescence points. These persons are candidates for Progenitors for Western European descendants. Most of the Progenitors will be included in your ancestry when you link your genealogy to the Unifying Ancestry.

Connectivity—Local Connectivity Metric

The own-cousin relationship was calculated based on the most recent ancestor with at least two ascents. Every person with multiple children who are ancestors of Prince George is a lineage coalescence point. The analysis ignores lineage coalescence points that are further back in time than the most recent lineage coalescence, reducing the number from 6700 to 2119. An implication is that multiple persons will have a common lineage coalescence point. The number of times a person is the own-cousin lineage coalescence point can be counted and used to form a local connectivity metric.

Local connectivity identifies locations in the genealogy where familial relationships are most dense. These persons in the Unifying Ancestry will typically have a large number of children. A local connectivity metric can be defined by counting the number of times (#Conn) a person is the lineage coalescence point for own cousin relationships. This identifies the persons that have the highest local connectivity within the genealogy. A metric which calculates this is:

- Local connectivity metric $= \mathrm{Log2(\#Conn)/(\#Gen)}$

where #Gen is the average number of generations from the own-cousin lineage coalescence point to the connected persons who have two ascents to the coalescence point. The average is calculated over all the lineages for which the person is the lineage coalescence point.

The persons listed in Table 10 identify the locations in the genealogy where there is the highest concentration of own-cousin lineage coalescence. Each person is a lineage coalescence point for multiple ancestors of the root of the genealogy. These persons will be expected to appear in lineages that link Western Europeans to the Unifying Ancestry. The column labeled Birth is the birth year. The column labeled Metric is the local connectivity metric. The column labeled #Gen is the average number of generations from the person to descendants that have two ascents to the person. The column labeled #Conn is the number of times the person is a lineage coalescence point. The column labeled #Ascents is the number of unique paths through the genealogy from Prince George of Cambridge to the person. The column labeled Relation is the relationship of the person to Charlemagne.

Noteworthy facts about Table 10 are:

- We would like to identify potential progenitors who are born before Charlemagne (born in 747 AD). Only four persons in the list qualify: Ansegisel of Brabant, Luitfried I Hugues d'Alsace, Recared I of Visigoths, and Bodegeisel II de Schelde.
- The person who is the own-cousin lineage coalescence point the largest number of times is Boleslav III of Poland with 208 connections.
- The persons with the largest local connectivity metric have #Gen values that are smaller than the average value found for all members of the unifying ancestry that have own-cousin relationships (7.45 generations).
- The number of ascents is not directly dependent on the number of times a person is a linage coalescence points. Boleslav III of Poland has only 3 million ascents even though he is the own-cousin lineage coalescence point 208 times.
- The four persons born before Charlemagne have lineages that depend upon cultural tradition.
- The person with the largest number of ascents in the list is Bodegeisel II de Schelde with 116 billion ascents from Prince George.
- Ten of the persons in the list are descendants of Charlemagne. Two are ancestors, and two are cousins of Charlemagne.

Ascents to Charlemagne will include ascents to his ancestors. This leaves two possible additional progenitors, Luitfried I Hugues d'Aslace and Recared I of Visigoths, both cousins of Charlemagne. Their common ancestors with Charlemagne are Dagobert I the Great, born about 605, and Clovis I of Franks, born about 466.

A better way to identify potential progenitors of Western Europeans is to look at own cousin lineage coalescence points that are not relatives of Charlemagne. The goal is to find unrelated progenitors that are born sufficiently far in the past that they can potentially be ancestors of all Western European descendants alive today.

Table 10 Persons with the largest local connectivity metric

	Person	Birth	#Gen	#Conn	Metric	#Ascents	Relation
1	Ansegisel of Brabant	602	4.75	55	1.2183	5.40E+10	2G-GFather
2	Luitfried I Hugues d'Alsace	704	5.54	89	1.169	4.05E+09	3C1R
3	Otto I of Guelph	1204	5.67	89	1.1413	3.56E+05	11G-GSon
4	Boleslav III of Poland	1086	6.78	208	1.136	3.16E+06	9G-GSon
5	Louis IV of Bavaria	1282	6.05	111	1.1223	5.76E+04	13G-GSon
6	Robert II Stewart	1316	6.49	148	1.1103	1866	15G GSon
7	Friedrich I von Brandenburg	1371	5.16	50	1.0938	9682	15G-GSon
8	Frederik I of Denmark	1471	5.67	67	1.0695	702	18G-GSon
9	Clementia of Poitiers	1058	7.06	170	1.0497	3.72E+06	8G-GDaughter
10	Jean II of France	1319	5.57	56	1.0423	5611	15G-GSon
11	Herbert II de Vermandoise	880	7.31	196	1.0422	7.72E+07	3G-GSon
12	Recared I of Visigoths	559	5.26	43	1.0324	7.50E+09	h2C6R
13	Bodegeisel II de Schelde	552	6.07	75	1.0267	1.16E+11	4G-GFather
14	Aubree of Lorraine	930	5.32	44	1.0266	2.33E+07	4G-GDaughter
15	Charlemagne the Great	747	6.34	89	1.0219	3.24E+09	Source

Birth is the Birth year

Metric is the local connectivity value, $\log2(\# Conn)/(\# Gen)$

#Gen is the average number of generations from the own cousin relationships to the lineage coalescence point

#Conn is the number of times the person is the lineage coalescence point

#Ascents is the number of ascents from Prince George of Cambridge

Relation is the relationship to Charlemagne

Research Genealogy—Completeness

Traditional metrics for the completeness of a genealogy are the identification of a royal descent or a descent from Charlemagne. As shown in Table 2, connecting to the Unifying Ancestry will lead to descents from hundreds of Kings. Table 11 lists the number of descents from Charlemagne to the current Kings and Queens of Europe. Because of the high connectivity of the Unifying Ancestry, the number of descents from Charlemagne to the current Kings and Queens of Europe is measured in the billions. Each descent is unique, with a different list of persons. The column labeled #Descents counts the number of unique paths from Charlemagne to the King or Queen. The column labeled #Persons counts the number of ancestors of the King or Queen who are descendants of Charlemagne.

Notable facts about the analysis include:

- All the current Kings and Queens have at least 11,000 ancestors who are descendants of Charlemagne.
- The number of descents is measured in the billions, with King Felipe VI of Spain having over 8 billion descents.
- Note that the largest number of descents from Charlemagne in the Research Genealogy is to Sophia of Wurttemberg with 10.49 billion descents.
- The number of descents from Charlemagne to Prince George of Cambridge is 3.24 billion through 19,313 persons. Even though the number of persons involved in the descents is measured in the thousands, the number of descents is measured in the billions because the lineages are highly intermarried.

Table 11 Number of descents from Charlemagne to the Kings and Queens of Western Europe

	Kings and Queens of Europe	#Descents (billion)	#Persons
1	Felipe VI of Spain	8.41	13,110
2	Margrethe II of Denmark	4.20	12,202
3	Carl XVI Gustav of Sweden	3.89	12,376
4	Charles Philip Mountbatten-Windsor	3.10	15,936
5	Harald V of Norway	2.87	11,995
6	Philippe Leopold Louis Marie of Belgium	2.43	15,207
7	Willem Alexander of the Netherlands	1.72	12,197

The Core Genealogy is the ancestors of George Alexander Louis Mountbatten-Windsor
#Descents is the number of descents from Charlemagne
#Persons is the number of ancestors of the King/Queen who are descendants of Charlemagne

In the Research Genealogy, Prince George of Cambridge has the largest number of ancestors (45,441) and the largest extended family (187,608 persons). Table 12 shows a distribution of the ancestors of Prince George of Cambridge by century, along with distributions for the Unifying Ancestry and the common ancestors of the Royal Families of Europe. The column labeled Year is the start of the century. The column labeled #People is the number of persons in the Research Genealogy born in the century. The columns labeled #Ancestor and #Com-core are the number of ancestors of Prince George of Cambridge born in the century. The column labeled #Com-anc-roy is the number of common ancestors of the Royal Families born in the century.

The distribution shows that the amount of information available as you go back in time dwindles to primarily notable historical persons who are common ancestors of the Royal Families of Western Europe. In the tenth century, the number of ancestors of Prince George (2255 persons) is only slightly larger than the number of common ancestors of the Royal Families (2230 persons). A Unifying Ancestry based on the ancestors of Prince George will include the historically important persons born before the tenth century. Connecting a genealogy to a Unifying Ancestry is a reasonable measure of Completeness that addresses the traditional measures of a royal descent or a descent from Charlemagne.

With the advent of DNA testing, an alternate measure of completeness is the demonstration that the ancestry of a person reflects the person's DNA analysis. To do this, a metric based on the ancestral distribution fraction was developed. This calculates the fraction of the ancestry that comes from each ancestor taking into account all ascents. The algorithm:

- Loops over the ancestors of the root person
- Assigns an ancestral fraction to each ancestor equal to half the sum of the ancestral fraction for each child.

Table 13 shows an analysis of the ancestral distribution fraction for Prince George of Cambridge. For ancestors with only one parent, half of the ancestral distribution fraction is assigned to their birth country. For ancestors with no parents, all the ancestral distribution fraction is assigned to their birth country. A viable genealogy will reproduce the expected country distribution found from a DNA analysis. The column labeled Percent identifies the fraction of the ancestry associated with each country. The column labeled Number identifies the number of ancestors who were born in the country.

Since the ancestors of Prince George of Cambridge comprise the Unifying Ancestry, the overlap with the Unifying Ancestry should be 100% instead of 93.9387%. The difference is because the birth location is missing for about 7% of the ancestors of Prince George. Note that 2.1132% of his ancestry is assigned to the United States, indicating that information about the ancestry of fifteen persons was incomplete, with lineages terminating in the United States.

Table 12 Distribution by century of the ancestors of Prince George of Cambridge

Year	#People	#Ancestor	#Com-core	#Com-anc-roy
0	2442	1473	1473	1473
100	384	321	321	321
200	389	272	272	271
300	527	373	373	368
400	824	496	496	492
500	912	501	501	499
600	936	507	507	503
700	1117	676	676	669
800	2021	1226	1226	1210
900	3534	2255	2255	2230
1000	6666	4212	4212	3794
1100	11,021	6207	6207	4615
1200	16,443	8097	8097	5007
1300	19,940	7572	7572	3472
1400	27,148	5872	5872	2110
1500	41,346	3084	3084	781
1600	66,507	1531	1531	271
1700	65,985	606	606	38
1800	41,259	142	142	0
1900	21,155	18	18	0
2000	54	1	0	0
Total	330,610	45,442	45,441	28,124

Year is the start of the century
Persons born before 0 AD are included in the first century
Persons with no birth year are included in the first century
#People is the number of persons whose birth date falls in the century
#Ancestor is number of ancestors of the root person
#Com-core is number of common ancestors with the core genealogy. The core ancestry is the ancestors of George Alexander Louis Mountbatten-Windsor
#Com-anc-roy is number of common ancestors with the common royal ancestors

The ancestral fractional distribution analysis of the Unifying Ancestry is significant because on average 99% of a person's ancestors comes from the Unifying Ancestry, but typically 1% of a person's ancestral distribution fraction comes from members of the Unifying Ancestry. The ancestral distribution fraction is dominated by the most recent

Table 13 Ancestral distribution fraction by century

Percent	Number	Country
62.33	4317	England
8.19	686	Scotland
7.48	992	Germany
3.23	304	Ireland
3.03	1417	France
2.11	15	United States
1.07	385	Italy
0.87	52	Poland
0.59	89	Russia
0.49	24	Hungary
93.94	9640	Total

Note that the birth country of Treetops is used
Number is the number of ancestors found in the country
For the Core Genealogy based on the ancestors of George Alexander Louis Mountbatten-Windsor
Total ancestral distribution fraction from Core Genealogy is 93.9387%

ancestors, but the connection to a national community is dominated by the Unifying Ancestry.

The concept of a link to the Unifying Ancestry replaces the traditional completeness metrics of a royal descent or a descent from Charlemagne. By connecting the genealogy to a Unifying Ancestry, it becomes possible to find familial relationships to every other person that is also able to link to the Unifying ancestry. A genealogy is complete when persons in the genealogy can be linked to all other members of their national community.

Progenitors for Western Europeans

If a set of progenitors exists for persons with Western European ancestry, it should be possible to find lineages for historically notable persons to the progenitors, provided they are also of Western European descent. This defines familial relationships between historically notable persons, creating a Genealogical History of the Modern World. Of interest is whether persons of non-Western European descent can be linked through multiple extended families to the same set of progenitors. This typically occurs through marriages of the descendants of notable persons.

The lists enumerating potential Progenitors of Western European descendants include:

- Persons alive in Europe in 1325   54 million

- Members of the Unifying Ancestry based on the ancestors of Prince   45,441
  George
- Common ancestors of the Royal Families of Europe   28,124
- Unifying Ancestry members with more than 1 billion ascents   2973
- Lineage coalescence points for own-cousin relationships in the Unifying   2506
  Ancestry

We can generate an even shorter list of progenitors by applying the following constraints to the list of lineage coalescence points:

- 2506 lineage coalescence points in the Unifying Ancestry
- 292 persons with at least 200 million ascents from Prince George of Cambridge
- 125 persons who are not an ancestor of Charlemagne
- 38 persons that are at least 56 generations from Prince George of Cambridge along a lineage of maximal ascent
- 27 persons that have at least 700 descendants distinct from the descendants of Charlemagne
- 18 persons born after 200 AD
- 18 persons with descendants born in the twentieth century
- 17 persons after excluding spouses
- 10 persons that are not ancestors of another candidate progenitor.

These constraints define a minimal list of 10 progenitors to which persons of Western European descent should be able to link their ancestry. The result is shown in Table 14, along with Charlemagne. Even though an explicit effort is made to find progenitors unrelated to Charlemagne, there is a strong overlap in the descendants of each progenitor with the descendants of Charlemagne. If your lineage connects to the Unifying Ancestry, you will automatically have ancestors in this list. Alternatively, you can look for a lineage to one of the progenitors, which then gives you a connection to the Unifying Ancestry.

In Table 14, Frac is the fraction of the relatives of Prince George of Cambridge who are descendants of the Progenitor. The column labeled Birth is the birth year. The column labeled #Desc is the number of descendants of the Progenitor. The column labeled #DescNC is the number of descendants who are not descendants of Charlemagne. The column labeled Ascents is the number of unique paths from Prince George of Cambridge to the progenitor. The column labeled Gen is the number of generations through a lineage of maximal ascent from Prince George to the progenitor. Lineages of Maximal Ascent are created by starting with the progenitor, selecting the child with the most ascents, and iterating down to the root person. The column labeled Period is the range of years during which the Progenitor has descendants not held in common with Charlemagne. The column labeled Top is set to one if the person does not have an own-cousin relationship.

Table 14 Progenitors for persons of Western European Descent

	Progenitor	Top	Frac	Birth	#Desc	#DescNC	Ascents	Gen	Country	Period
	Charlemagne the Great	0	0.805	747	150,942		3.20E+09	49	France	
1	Potitus Photaighe of Ireland	1	0.811	300	152,209	1339	1.20E+11	64	Ireland	325–1742
2	Eudes of the Gewissi	0	0.785	320	147,316	1426	2.50E-10	62	England	342–1814
3	Magnus Clemens Maximus	1	0.784	340	147,152	1262	1.60E-10	60	Spain	360–1814
4	Coel Hen	0	0.785	380	147,212	1322	8.40E-09	59	England	409–1814
5	Danp of Denmark	1	0.81	384	151,887	2209	1.30E+10	63	Denmark	404–1930
6	Vortigern of Powys	1	0.784	395	147,133	1243	9.40E+09	59	Wales	413–1814
7	Pompeius of Dyrrhachium	1	0.804	400	150,906	1333	1.50E+10	57	Albania	430–1963
8	Cunedda the Great	0	0.784	409	147,085	1195	4.40E+09	58	Scotland	425–1590
9	Clodgar von Therouanne	0	0.811	445	152,202	1332	5.90E+10	59	France	480–1742
10	Charibert II Caribert of Neustria	0	0.811	565	152,131	1206	5.30E+10	57	France	585–1742

Top is set to one if the progenitor does not have an own cousin relationship
Frac is the fraction of the relatives of Prince George who are descendants of the progenitor
Birth is the birth year (negative numbers are before 0 AD)
#Desc is the number of descendants of the progenitor
#DesNC is the number of descendants who are not descendants of Charlemagne
Ascents is the number of ascents from Prince George of Cambridge
Gen is the number of generations from the Prince George along a lineage of maximal ascent
Country is the birth country
Period is the range in years with descendants not held in common with Charlemagne

The notable facts inferred from Table 14 include:

- The descendants of each progenitor strongly overlap with the descendants of Charlemagne.
- Typically, each progenitor has only 1000–2200 additional descendants different from the Charlemagne descendants.
- To show that the additional descendants exist to the present day, the range of birth years for the additional descendants is printed in the column labeled "Period". Progenitors are significant if they have descendants different from the descendants of Charlemagne in the nineteenth century.
- More than 78% of the relatives of Prince George are descendants of each Progenitor.
- Potitus Photaighe of Ireland claims the largest number of relatives of Prince George of Cambridge as descendants, while Charlemagne does nearly as well (81.1–80.5%).
- All the progenitors have more than three billion ascents from Prince George of Cambridge.
- The common descendants of the 10 progenitors listed in Table 14 are also descendants of Charlemagne.
- All the progenitors were born in Europe.

The multiple nationalities that are traversed by the descendants of the Progenitors can be illustrated by listing lineages of Maximal Descent from a progenitor to Prince George of Cambridge. A lineage of Maximal Descent is generated by the steps:

- Find the number of descents from the progenitor to each descendant.
- Starting with Prince George, identify the parent with the largest number of descents.
- Track the lineage through the selected parent and iterate until the Progenitor is reached.

Table 15 gives the lineage of Maximal Descent from Danp of Denmark to Prince George of Cambridge. Note that information about Danp of Denmark comes from the Heimskringla Saga. In Table 15, the column labeled Descents is the number of unique paths from Danp of Denmark to the descendant. The column labeled Relation is the familial relationship of the descendant to their spouse.

Table 15 Lineage of maximal descent from Danp of Denmark to Prince George of Cambridge

Gen	Person	Birth	Death	Mar	Descents	Relation
0	George Mountbatten-Windsor	2013			1.28E+10	
1	William Mountbatten-Windsor	1982		2011	1.28E+10	13C1R
2	Charles Mountbatten-Windsor	1948		1981	1.24E+10	7C1R*
3	Philip Mountbatten	1921	2021	1947	8.11E+09	3C
4	Victoria Alice of Battenberg	1885	1969	1903	4.45E+09	4C
5	Victoria of Hesse and the Rhine	1863	1950	1884	3.43E+09	1C1R*
6	Ludwig of Hesse and the Rhine	1837	1892	1862	1.78E+09	4C
7	Karl of Hesse and the Rhine	1809	1877	1836	1.02E+09	2C
8	Whilhelmine Louise of Baden	1788	1836	1804	5.51E+08	1C
9	Karl Ludwig of Baden	1755	1801	1774	3.34E+08	1C
10	Karl Friedrich of Baden	1728	1811	1751	1.88E+08	h3C1R
11	Anna of Orange-Nassau	1710	1777	1727	9.49E+07	4C
12	Marie Louise of Hesse-Cassel	1688	1765	1709	5.54E+07	h3C
13	Karl of Hesse-Cassel	1654	1730	1673	3.07E+07	1C
14	Hedwig Sophie of Brandenburg	1623	1683	1649	1.69E+07	2C
15	George William of Brandenburg	1595	1640	1616	1.07E+07	4C1R
16	Anna of Prussia	1576	1625	1594	5.77E+06	3C1R
17	Mary Eleanor of Cleves-Julich	1550	1608	1573	3.48E+06	2C1R*
18	Maria of Austria	1531	1581	1546	1.93E+06	4C
19	Ferdinand I of Austria	1503	1564	1521	1.25E+06	h2C1R*
20	Philip I of Austria	1478	1506	1496	8.72E+05	3C1R
21	Mary of Burgundy	1457	1482	1477	6.37E+05	2C
22	Isabella of Bourbon	1436	1465	1454	3.47E+05	1C
23	Agnes of Burgundy	1407	1476	1425	2.08E+05	2C
24	Jean of Burgundy	1371	1419	1385	1.31E+05	3C
25	Margaret II of Flanders	1350	1405	1369	8.37E+04	3C
26	Margaret of Brabant	1323	1368	1347	4.27E+04	h3C1R
27	Marie d'Evreux	1302	1335	1314	2.44E+04	2C
28	Marguerite d'Artois	1285	1311	1301	1.28E+04	2C1R*
29	Blanche of Brittany	1270	1327	1281	6.77E+03	4C1R
30	John II of Brittany	1239	1305	1260	4.29E+03	h4C2R
31	John I of Brittany	1217	1286	1236	2.40E+03	3C1R*

(continued)

Table 15 (continued)

Gen	Person	Birth	Death	Mar	Descents	Relation
32	Peter I de Braine of Dreux	1190	1250	1213	1.53E+03	4C
33	Yolande de Coucy	1168	1222	1184	1.13E+03	3C1R*
34	Agnes of Hainault	1142	1170	1164	8.86E+02	4C1R*
35	Baldwin IV of Hainault	1108	1171	1130	5.70E+02	h1C
36	Baldwin III of Hainault	1088	1120	1107	3.62E+02	5C1R
37	Baldwin II of Hainault	1056	1098	1084	2.56E+02	3C
38	Richildis von Mons	1030	1087	1055	1.42E+02	2C
39	Mathilde of Verdon	994	1051	1015	8.20E+01	4C
40	Mathilde de Dagsbourg	977			5.00E+01	3C2R*
41	Judith d'Oehningen	961	1022		5.00E+01	
42	Richilde of Saxony	946	1014		4.70E+01	3C
43	Liudolf of Saxony	930	957	947	3.50E+01	4C
44	Otto I the Great	912	973	929	2.70E+01	5C1R
45	Matilda von Ringleheim	896	968	909	1.80E+01	3C1R*
46	Ragnhildis von Friesland	868	917	900	1.50E+01	4C
47	Godofrid of Haithabu	829	885	882	1.50E+01	4C2R*
48	Sigrid of Westfold	810	854		8.00E+00	1C2R*
49	Aslaug of Ringerike	791			5.00E+00	7C1R
50	Helen of England	776			4.00E+00	h3C3R*
51	Elia of Northumberland	761			4.00E+00	
52	Eahlmund of Kent	746	796		2.00E+00	1C1R
53	Etherburge of Kent	731			2.00E+00	18C3R
54	Melli von Friesland	716			2.00E+00	h10C
55	Radbodus III von Friesland	699	768		2.00E+00	
56	Adgillis III von Friesland	682			2.00E+00	
57	(daughter) von Gardarige	665			2.00E+00	
58	Roric Slingeband	650	720		1.00E+00	h4C2R*
59	Halfdan Berg-Gram Frodisson	627			1.00E+00	
60	Frodi Hroereksson	590			1.00E+00	
61	Hroerek Ringniggard	553			1.00E+00	
62	Ingjald Frodesson	516			1.00E+00	
63	Frode VII the Valiant	479	548		1.00E+00	

(continued)

Table 15 (continued)

Gen	Person	Birth	Death	Mar	Descents	Relation
64	Fridleif Frodisson	456			1.00E+00	
65	Frodi Mikillati	433			1.00E+00	
66	Dan the Proud	404			1.00E+00	
67	Danp of Denmark	384			0.00E+00	

In Table 15:

Gen is the number of generations in the lineage of maximal descent back to Danp
Birth, Death, and Mar are birth, death, and marriage years
Descents is the number of descents from Danp of Denmark
Relation is the relationship between the spouses when known

The lineage of Maximal Descent from Danp of Denmark shown in Table 15 does provide useful information about descents from Progenitors:

- At each generation the spouses are usually related, with cousin relationships varying between 1st cousin and 10th cousin. The most distantly related spouses in the list are Prince George's parents at 13th cousin once removed.
- The spousal relationship was not found for 14 of the persons in the list.
- Prince George of Cambridge has 12.8 billion descents from Danp of Denmark. The number of descents first exceeds 1 million in the 19th generation with Ferdinand I of Austria.
- The lineage of Maximal Descent traverses 67 generations, which is greater than the analytic estimate based on population genetics for the number of generations to progenitors (49).

Other lineages can be derived for tracking connectivity that represent:

- Lineage of Maximal Ascent. At each generation starting with Danp of Denmark, the child with the largest number of ascents is selected.
- Lineage of Eldest Descent. At each generation starting with Danp of Denmark, the child born first that is an ancestor is selected.
- Lineage of Youngest Descent. At each generation starting with Danp of Denmark, the child born last that is an ancestor is selected.
- Lineage with the smallest number of generations from Danp of Denmark to Prince George.

In Appendix B, Tables B.1–B.10 provide lineages of Maximal Ascent from Prince George of Cambridge to each progenitor listed in Table 14. These lineages traverse quite different paths through the Research Genealogy than the lineage of Maximal Descent. For Danp of Denmark, the lineage of Maximal Descent has 67 generations, while the lineage of Maximal Ascent has 63 generations.

Research Genealogy—Coherence

A demonstration of coherence requires linking the knowledge base (constructed by evaluating the properties of Consistency, Correctness, Closure, Connectivity, and Completeness) to a community consensus knowledge base. This ensures that the events used to generate the genealogy can be tied to historical events. If the descendants from the list of progenitors include historically notable persons, we will have demonstrated that the familial relationships derived in the Unifying Ancestry are compatible with the familial relationships between historically notable persons.

Essential Information Metric

Coherence also requires that the knowledge base includes all essential information needed to demonstrate a unifying ancestry and identify the existence of progenitors for a national community. Since all members of a National Community will have lineages to the Unifying Ancestry, the ancestors of any member could be chosen as the Unifying Ancestry. An essential information metric provides a way to choose the best Unifying Ancestry from all the candidates.

One way to analyze genealogies for whether essential information is missing is to model the average number of ascents and compare the predicted number with the observed number. We need a metric that calculates the average number of ascents across the entire genealogy for each generation. We can use the own cousin relationship analysis to do this. This analyzes the average number of times lineages combine at each lineage coalescence point (c), and the average number of generations to a lineage coalescence point (g). The expected average number of ascents to persons within each generation is then given by

$$A = c^{(n/g)} \text{ where n is the generation number} \tag{3}$$

For a progenitor, the number of generations that lineages need to be traced is proportional to $\log 2(P)$ where P is the population size.

$$n = b * \log 2(P) \tag{4}$$

We can solve for the value of "b" by requiring that the average number of ascents to a valid progenitor be at least the population size since each member of the population has an ascent to the progenitor through the Unifying Ancestry.

The expected number of ascents is then given by

$$A_c = 2^{(Log2(c)\ Log2(P)\ b/g)}$$

This can be rewritten as:

$$A_c = P^{(b\ Log2(c)/g)} \tag{5}$$

For the average number of ascents to exceed the population size, the value of the exponent on the population needs to be at least 1. This defines a metric for whether essential information is missing from the genealogy. We require

$$X = b\ Log2(c)/g > 1 \tag{6}$$

If we set

$$b = g/Log2(c)$$

the value of the exponent will be at least one. The minimum number of generations to a progenitor, based on the information content of the genealogy is then given by

$$N_e = g\ Log2(P)/Log2(c) \tag{7}$$

Equation (7) is a generalization of Eq. (2) that takes into account the information actually present within the genealogy to determine when in the past progenitors can be found. Given that every person in a national community is linked to the unifying ancestry, the ancestors of any person can be used as the unifying ancestry. The ancestry that has the smallest number of generations to a progenitor is the preferred unifying ancestry.

Table 16 evaluates the "essential information metric" for prior versions of the Research Genealogy. We see that as the number of common Royal ancestors increases, the "essential information metric" slowly increases. An extrapolation predicts that roughly 32,000 common ancestors for the royal families of Western Europe will need to be identified to ensure that relevant information has been included. Conversely, we can extend the number of generations back further in time to account for missing information and this way raise the missing information exponent value to 1. This gives a genealogy-based estimate for the number of generations to the time when everyone alive is a progenitor:

The progenitors listed in Table 14 have lineages of maximal ascent that traverse 55–63 generations and are thus viable candidates for progenitors for all Western Europeans alive today.

Table 16 Analysis of missing essential information

Size	#Common	g	c	X	Ng
279,887	18,408	8.49	14.3	0.8	61.0
294,383	18,732	8.46	14.56	0.808	60.4
312,841	19,684	8.46	14.8	0.813	60.0
324,355	23,756	8.33	14.8	0.826	59.1
330,610	28,409	7.81	14.57	0.876	55.7
330,610*	28,124	7.45	16.03	0.951	51.3
337,121	33,131	7.29	15.81	0.967	50.5

Size number of persons in the Research Genealogy version
#Common number of common ancestors of the Kings and Queens
g average #generations to a lineage coalescence point
c average number of lineages that coalesce
X Essential information metric $= 1.77 \, \text{Log2}(c)/g$
Ng number of generations to the time when everyone is a progenitor
Population size is assumed to be 200 million

If the analysis is restricted to just the common ancestors of the Royal Families of Western Europe, an even smaller number of generations to progenitors is found, as shown in the bottom two lines of Table 16. This shows that the common ancestors of the Royal Families have a higher degree of connectivity (more lineage coalescence) than all the ancestors of Prince George of Cambridge.

For genealogies that have no missing information, the number of generations to trace ancestors to a progenitor, N_e, is 49 for a population size of 200 million persons.

Historically Notable Persons

We can test whether the list of progenitors in Table 14 is significant by checking whether lineages exist to the persons in these test cases.

- Notable person—Barack Obama.
- Group of persons—U. S. Presidents.
- Historically notable persons—identified in Hart's "100 most influential persons in history" (Hart, 1992)
- Historically notable persons—identified in Palmer's History of the Modern World, which includes twelve of the U. S. Presidents and 48 of the persons in Hart's list.

The check proceeds in multiple steps:

- Identify whether the persons in the test case have a set of common ancestors
- Identify the Progenitors for the common ancestors
- Verify that the Progenitors listed in Table 14 are included in the common ancestors of the persons in the test case.

The number of ancestors in the Research Genealogy for each U. S. Presidents is listed in Table 17. The column labeled Rel is the relationship of the U. S. President to George Washington. The column labeled #Anc is the number of ancestors of the U. S. President, and the column labeled #CoreAnc is the number of these ancestors who are also ancestors of Prince George of Cambridge.

As shown in Table 17, Barack Obama has 22,015 ancestors in the Research Genealogy, of which 20,468 are members of the Unifying Ancestry based on the ancestors of Prince George of Cambridge. Barack Obama is a descendant of Charlemagne and the 10 progenitors listed in Table 14.

Table 17 also shows that the U. S. Presidents are all related to George Washington. The cousin relationships vary from 3rd cousin 8 removes* for Barack Obama to 12th cousin 5 removes* for Dwight Eisenhower.

The Research Genealogy contains between 10,271 ancestors for Dwight Eisenhower to 25,550 ancestors for George Walker Bush. The table also lists the number of ancestors for each President that are members of the Unifying Ancestry based on the ancestors of Prince George of Cambridge.

- On average, 96% of each President's ancestors are in the Unifying Ancestry.
- The number of common ancestors of the U. S. Presidents in the Research Genealogy is 8072 persons.
- All the common ancestors of the U. S. Presidents are members of the Unifying Ancestry based on the ancestors of Prince George of Cambridge.
- All the U. S. Presidents are descendants of the Progenitors listed in Table 14.

The common ancestors of the U. S. Presidents can be analyzed to find a set of presidential Progenitors, the persons who together with their ancestors comprise the common ancestors. Table 18 lists 43 persons who are the presidential progenitors for the U. S. Presidents. Presidential progenitors are found across multiple national communities, from England, to Spain, to France, to Scandinavia. The column labeled ID is the unique person identifier used for the person in the Research Genealogy. The columns labeled Birth and Death provide the birth and death years for the person.

The most recent common ancestor is Blanche de Brienne, born in 1251. The number of generations from Prince George of Cambridge to Blanche de Brienne is 23. Note that Edward I Plantagenet, King of England, is also a common ancestor of the U. S. Presidents. The number of generations from Prince George to Haakon Sigurdsson, born is 951, is 30.

Table 17 The relationship of the U. S. Presidents to George Washington

	Person	Rel	#Anc	#CoreAnc
1	George Washington	Source	16,454	15,851
2	John Adams	8C 3R*	15,284	15,111
3	Thomas Jefferson	9C 1R*	12,269	12,110
4	James Madison	10C 2R*	14,707	14,421
5	James Monroe	11C 3R*	12,924	12,836
6	John Quincy Adams	8C 4R*	15,569	15,232
7	Andrew Jackson	H 7C 3R*	14,427	14,332
8	Martin van Buren	11C 2R*	12,162	12,060
9	William Henry Harrison	10C	13,406	13,207
10	John IV Tyler	10C 3R*	12,710	12,548
11	James Knox Polk	11C 3R*	12,832	12,703
12	Zachary Taylor	9C	14,826	14,555
13	Millard Fillmore	10C 4R*	12,782	12,374
14	Franklin Pierce	9C	14,736	14,326
15	James Buchanan	11C 2R*	10,848	10,748
16	Abraham Lincoln	8C 4R*	16,216	15,776
17	Andrew Johnson	10C 4R*	11,538	11,443
18	Ulysses Simpson Grant	6C 2R*	16,830	16,285
19	Rutherford Birchard Hayes	7C 3R*	15,867	15,124
20	James Abram Garfield	8C 4R*	14,752	13,974
21	Chester Alan Arthur	10C 3R*	13,381	13,122
22	Stephen Grover Cleveland	7C 5R*	16,284	15,805
23	Benjamin Harrison	10C 2R*	14,343	13,869
24	Stephen Grover Cleveland	7C 5R*	16,284	15,805
25	William McKinley	8C 4R*	16,663	16,560
26	Theodore Roosevelt	8C 6R*	16,532	16,143
27	William Howard Taft	6C 5R*	19,418	18,205
28	Thomas Woodrow Wilson	11C 3R*	10,574	10,539
29	Warren Gamaliel Harding	8C 5R*	14,996	14,676
30	John Calvin Coolidge	6C 6R*	18,663	16,926
31	Herbert Clark Hoover	9C 7R*	15,289	14,556
32	Franklin Delano Roosevelt	8C 4R*	20,700	19,473

(continued)

Table 17   (continued)

	Person	Rel	#Anc	#CoreAnc
33	Harry S. Truman	9C 6R*	14,195	13,892
34	Dwight David Eisenhower	12C 5R*	10,271	10,093
35	John Fitzgerald Kennedy	11C 7R*	12,218	12,144
36	Lyndon Baines Johnson	9C 7R*	13,720	13,531
37	Richard Milhous Nixon	8C 7R*	17,242	16,330
38	Gerald Rudolph Ford	8C 7R*	18,009	16,620
39	James Earl Carter	H 9C 6R*	16,366	15,705
40	Ronald Wilson Reagan	8C 8R*	17,485	17,405
41	George Herbert Walker Bush	7C 8R*	23,223	20,240
42	William Jefferson Clinton	6C 9R*	20,439	20,073
43	George Walker Bush	7C 9R*	25,550	21,143
44	Barack Hussein Obama	3C 8R*	22,015	20,468
45	Donald John Trump	11C 8R*	12,961	12,836
46	Joseph Robinette Biden	11C 8R*	12,087	11,986

All the Presidential Progenitors are descendants of Charlemagne except for Alvidis (de Broyes and Beaufort), Arne Arnmodsson and his spouse, Thora Thorstein's-dottir, Geoffrey II of Semur, and Haakon Sigurdsson and his spouse (concubine of Norway). All but eleven of the Presidential Progenitors are descendants of all the progenitors listed in Table 14. The exceptions are:

Hugh VIII de Lusignan, descendant of Danp, Pompeius, Potitus, Charibert II and Clodgar
Louis II von Loos, descendant of Danp, Potitus, Charibert II, and Clodgar
Elisabeth de Namur, descendant of Danp, Pompeius, Potitus, Charibert II, and Clodgar
Erneburga (d'Estouteville), spouse of Robert II d'Estouteville
Edith de Warenne, descendant of Danp, Potitus, Charibert II, and Clodgar
Agnes de Mortain, descendant of Danp, Potitus, Charibert II, and Clodgar
Melisende de Montlhery, descendant of Danp, Pompeius, Potitus, Charibert II, Clodgar
Hugues I Bardoul de Broyes and Beaufort, descendant of Danp, Pompeius, Potitus, Charibert II, and Clodgar
Alvidis (de Broyes and Beaufort), spouse of Hugues I Bardoul de Broyes and Beaufort
Thora Thorstein's-dottir, spouse of Arne Arnmodsson
Arne Arnmodsson
Geoffrey II of Semur
Concubine of Norway, spouse of Haakon Sigurdsson

Table 18 Ancestors of 43 persons comprise the common ancestors of the U. S. Presidents

	Presidential progenitor	ID	Birth	Death
1	Blanche de Brienne	8541	1251	
2	Blanche d'Artois	6954	1248	1302
3	William II de Fiennes	8542	1244	1302
4	Eleanor of Castile	4107	1240	1290
5	Edward I Plantagenet	4104	1239	1307
6	Beatrice of Savoy	11,721	1213	1258
7	Isabel Le Bigod	7328	1212	
8	Alan MacDonal	5499	1186	1234
9	Margaret de Braose	8991	1178	
10	Walter de Lacy	8990	1168	1241
11	Yolande de Coucy	15,403	1168	1222
12	Bertrade de Montfort	3783	1155	1227
13	Robert II of Dreux	14,394	1154	1218
14	Hugh of Kevelioc	3782	1147	1181
15	Margaret de Huntingdon	5470	1146	1201
16	Petronilla de Grandmesnil	5732	1134	1212
17	Robert III de Beaumont	2968	1130	1190
18	Bourgogne de Rancon	13,955	1130	1169
19	Mathieu I of Lorraine	11,016	1125	1176
20	Agnes of Metz	40,088	1122	1175
21	Bertha Hohenstaufen	14,342	1121	1194
22	Hugh VIII de Lusignan	11,402	1107	1173
23	Helvide de Baudemont	15,008	1105	1165
24	Guy I de Dampierre	15,007	1100	1151
25	Louis II von Loos	40,089	1100	1191
26	Elisabeth de Namur	14,807	1095	1160
27	Erneburga (d'Estouteville)	108,990	1093	
28	Robert II d'Estouteville	108,989	1085	1140
29	Maud of Brittany	10,870	1084	1135
30	Walter de Gant	6892	1080	1138
31	Edith de Warenne	13,946	1078	
32	Agnes de Mortain	7303	1069	

(continued)

Table 18 (continued)

	Presidential progenitor	ID	Birth	Death
33	Ermengarde of Anjou	11,403	1068	1147
34	Alan IV Fergant of Brittany	5476	1066	1119
35	Andre I de Vitre	7302	1057	
36	Melisende de Montlhery	18,490	1042	
37	Hugues I Bardoul de Broyes and Beaufort	43,001	1016	
38	Alvidis (de Broyes and Beaufort)	94,540	1015	
39	Thora Thorstein's-Dottir	18,206	972	1015
40	Arne Arnmodsson	10,654	965	1024
41	Geoffrey II of Semur	41,386	961	1019
42	(concubine) of Norway	18,162	952	
43	Haakon Sigurdsson	10,539	951	994

Haakon Sigurdsson, descendant of Danp

The five exceptions that do not connect to a progenitor for Western Europeans were born before 1016 AD and had less than 24 generations to connect to Potitus Photaighe of Ireland.

References

Chang, J. T. (1999). Recent common ancestors of all present-day individuals. *Northern Ireland: Advances in Applied Probability, 31*, 1002–1026.

Hansen, C. M., & Thompson, N. D. (2012). *Ancestry of Charles II, King of England*. American Society of Genealogy.

Hart, M. H. (1992). *The 100: A ranking to the most influential persons in history*. Hart Publishing company.

McNaughton, A. (1973). *The book of Kings: A Royal Genealogy*. Gladstone Press.

Palmer, R. R., & Colton, J. (1952). *A history of the modern world*. Alfred A Knopf.

Runciman, S. (1962). *A history of the Crusades*. Cambridge University Press.

Schwinnicke, D. (1995). *Europaische Stammtafeln: Stammtafeln zur Geschichte der Europaischen Staaten, Neue Folge*. J. A. Stargardt.

Sturluson, S. (1964). *Heimskringla, history of the kings of Norway*. University of Texas Press.

Weis, F. L. (1982). *The Magna Charta Sureties*. Genealogical Publishing Co., Inc.

Genealogical History of the Modern World

The demonstration that the Unifying Ancestry knowledge base can be linked to a community-consensus knowledge base such as the common ancestors of the U. S. Presidents was relatively easy. Linking to historical events is more difficult. Since historical events are used to identify information about individuals, Palmer's History of the Modern World has been chosen as the knowledge base that represents the physical world. Historical events span the known world, including both Western European and Eastern European cultures. In addition, the timeline shown in Table 2.4 for persons in the Research Genealogy goes from Ancient Greece to the present. For many historically notable persons there is limited information about their lineages, making it difficult to link their ancestry.

Palmer's History of the Modern World identifies 719 notable persons. For each notable person, a lineage was sought that would link the person to the Unifying Ancestry. Table 1 provides a distribution by century of these notable persons. The column labeled Persons is the number of the notable persons born in the century. The column labeled Ancestor is the number of notables who are ancestors of Barack Obama and the column labeled CoreAnc is the number of notables who are members of the Unifying Ancestry based on the ancestors of Prince George of Cambridge.

A total of 604 of the notable historical persons have been linked into the Research Genealogy. The CoreGen3 analysis shows that:

- 13 of the notable persons are ancestors of Barack Obama.
- 46 of the notable persons are members of the Unifying Ancestry and ancestors of Prince George.
- 264 of the notable persons are cousins of Prince George and have lineages directly linking them to persons in the Unifying Ancestry.
- 294 of the notable persons are linked through multiple extended families.

© The Author(s), under exclusive license to Springer Nature Switzerland AG 2023
R. W. Moore, *Trustworthy Communications and Complete Genealogies*, Synthesis Lectures on Information Concepts, Retrieval, and Services,
https://doi.org/10.1007/978-3-031-16836-9_4

Table 1 Persons identified in
Palmer's history of the modern
world

Cent	Persons	Ancestor	CoreAnc
0	4	0	0
100	0	0	0
200	0	0	0
300	2	1	1
400	1	1	1
500	0	0	0
600	0	0	0
700	2	1	1
800	0	0	0
900	3	1	1
1000	4	2	2
1100	6	3	3
1200	6	2	2
1300	6	1	1
1400	41	0	8
1500	65	1	9
1600	74	0	9
1700	171	0	7
1800	212	0	1
1900	7	0	0
2000	0	0	0
Ave	1688	1026	1441
Tot	604	13	46

304S: Book, Palmer, R.R. and Joel Colton, A History of the Modern World, Alfred A Knopf, Change Date 25 JAN 2009
The number of ancestors is found for the root person Barack Hussein Obama Jr
The core genealogy is based on the ancestors of George Alexander Louis Mountbatten-Windsor. Persons references all persons in the genealogy. Ancestor references ancestors of the root person. CoreAnc references persons in the core genealogy
Ave. is the mean birth year averaged across all referenced persons
Tot. is the total number of persons summed across all centuries
Source 304, distribution of total number of persons referenced per century

Table 2 Connections to notable historical persons from the Core Genealogy

Degree of separation	# persons	# non-Western European	Percent (%)
0	310	71	23
1	152	29	19
2	80	20	25
3	33	10	30
4	18	8	44
5	3	2	67
6	4	2	50
7	3	3	100
8	1	1	100

- Appendix C provides 27 sample lineages of Maximal Ascent from a historically notable person to Charlemagne. The selected persons appear in both Palmer's "History of the Modern World" and Hart's "The 100".

Table 2 shows the number of historically notable persons present in each ring of extended families about the initial extended family for Prince George of Cambridge. The column labeled # persons is the number of notables present within a given degree of separation. The column labeled # non-Western European" is the number of notables not born in Western Europe. The column labeled Percent is the ratio of the # non-Western European and the # persons.

Palmers' History of the Modern World includes persons born in Eastern Europe, North and South America, and Asia. In the 4th through 8th rings around the extended family for the Unifying Ancestry, 55% are born outside Western Europe. For instance, the person in the 8th ring is Liu Shaoqi of China. The persons in the 7th ring are Mao Tse-Tung of China, Bela Kun of Romania, and Imre Nagy of Hungary.

On-going research is investigating how to link the remaining 115 notable persons to the Research Genealogy. The expectation is that if trustworthy information sources can be identified, lineages will be found for many of the missing persons. The effort required to link each person depends upon how far back in time lineages must be traced to connect their ancestry to the Unifying Ancestry. Table 3 shows the average connection year to the Unifying Ancestry for persons born in each century.

In Table 3, the column labeled Year is the start of the century. Persons born before 0 AD are included in the total for the first century. Also, persons with no birth year are included in the first century. The column labeled #core is the number of ancestors of Prince George of Cambridge born in the century. The column labeled #rel is the number of relatives of Prince George that are born in the century. The column labeled AveYear is the average connection year for persons born in the century. The column labeled AveGen

Table 3 Average connection year distribution by century

Year	#core	#rel	AveYear	AveGen
0	1473	614	−952.9	4.41
100	321	41	−100.3	4.95
200	272	80	95.5	3.75
300	373	117	230.2	3.76
400	496	278	245.6	4.28
500	501	343	282.4	5.32
600	507	385	473.6	5.01
700	676	372	627.4	4.16
800	1226	637	756	3.37
900	2255	949	900.6	2.36
1000	4212	1711	1018.7	2.11
1100	6207	3255	1121.1	1.97
1200	8097	5182	1225.9	1.94
1300	7572	6886	1308.5	2.39
1400	5872	10507	1401	2.75
1500	3084	17398	1476.3	3.55
1600	1531	28681	1526.2	5.22
1700	606	33477	1555.8	7.25
1800	142	18808	1578.3	9.57
1900	18	12410	1622.6	11.59
2000	1	36	1713.3	10.19
Total	45442	142167	1464.9	6.04

Year is the start of the century
Persons born before 0 AD are included in the first century
Persons with no birth year are included in the first century
#core is the number of core members whose birth date falls in the century
#rel is number of relatives whose birth date falls in the century
AveYear is average connection year for relatives born in the century
AveGen is the number of generations for the relatives to connect

is the average number of generations for persons born in the century to connect to the ancestors of Prince George.

A total of 12,410 connections were analyzed to generate the connection year for persons born in the twentieth century. Note that in the 1600s there are 1531 persons in the Unifying Ancestry. The maximum number of persons found in a century in the Unifying

Ancestry occurs in the thirteenth century with 8097 persons. Historically notable persons born before 700 AD are harder to connect as the average number of generations for connecting relatives increases to more than 5.

One of the original motivations for constructing the Research Genealogy and development of the genealogy analysis workbench was to verify a family tradition that we were related to Queen Elizabeth. This is now documented in the Research Genealogy as 10C1R* using a non-Royal lineage. This is consistent with Table 3. For relatives of Queen Elizabeth born in the twentieth century, on average their ancestry needs to be traced back to 1622, or about 11 generations, to connect to the Unifying Ancestry.

Historically Influential Persons

A list of the 100 most influential persons in history has been published by Michael Hart (Hart, 1992). All the persons of Western European descent in the list have been linked to the Research Genealogy. Lineages were found for 45 persons that linked them to the Unifying Ancestry. Appendix C provides lineages of maximal ascent from 27 of the influential persons to Charlemagne. All the persons in Appendix C are descendants of Charlemagne. They are also descendants of all the progenitors listed in Table 3.14 except Odo of Chatillon (known as Pope Urban II). This illustrates the comprehensive overlap of descendants between Charlemagne and the Progenitors listed in Table 3.14.

The time span for birth years ranges from 1035 to 1874. Of the notable persons, Odo of Chatillon was born the earliest, in 1035. Odo of Chatillon is a descendant of the progenitors Potitus of Ireland, Danp of Denmark, Clodgar von Therouanne, and Charibert II of Neustria

Reference

Hart, M. H. (1992). *The 100: A ranking to the most influential persons in history.* Hart Publishing company.

Trustworthy Communications

Communications are trustworthy when their information content can be related to a community knowledge base. This makes it possible to interpret the communication content in a manner consistent with the originator's intent. If a community knowledge base is not available, it is still possible to evaluate the properties of information extracted from the communications. The approach requires the following steps:

- Identify the information elements that will be extracted from the communications.
- Associate a list of sources and the communication dates with each information element.
- Identify the types of relationships that will be established between the sources. Examples include temporal relationships, spatial relationships, and community membership.
- Build a graph database linking the relationships between the sources for each information element.
- Analyze the properties of the database for Consistency, Correctness, Connectivity, Closure, Completeness, and Coherence. This will require defining metrics for evaluating each property that are relevant to the chosen relationships.
- Identify progenitors (primary source) for each information element.
- Identify influencers (sources linking the largest number of information elements).
- Identify lineages that track the flow of information between sources.
- Identify communities associated with each flow of information.
- Partition the information by community.
- Link the communities to a knowledge base that tracks community properties.
- Evaluate information by the properties of each community defined in the external knowledge base.

Instead of comparing the information content to a knowledge base appropriate for the domain of discourse, the information can be partitioned into originating communities.

R. W. Moore, *Trustworthy Communications and Complete Genealogies*, Synthesis Lectures on Information Concepts, Retrieval, and Services, https://doi.org/10.1007/978-3-031-16836-9_5

Trustworthiness is then assigned based on the community that generates the information flow and its ranking in the knowledge base that tracks properties of communities.

This approach utilizes many of the same analyses that were developed for trustworthy genealogies.

- Consistency is checked to ensure that every information element has an associated list of sources and communication dates.
- Closure is checked to see that every source is a member of a community.
- Correctness is checked to see that no source references occur before the information is available.
- Connectivity is evaluated to identify progenitors (global connectivity) and influencers (local connectivity).
- Completeness is checked to see if all communities are represented in the external knowledge base.
- Coherence is checked by verifying source membership in the dominant source community providing the information.

The resulting database categorizes the communications, identifies which information elements are local to a specific community, and which information elements are globally accepted across all communities. The approach is relevant whether the communications come from the past or are part of present-day discourse.

Genealogy Communications

Each source used to identify information about persons in a genealogy can be thought of as a communication from the past. The sources can be analyzed for trustworthiness by grouping them into categories and evaluating the trustworthiness of each category. If the categories are trustworthy, then the extracted lineages are assumed to be trustworthy.

In Appendix B (Progenitor lineages) and Appendix C (Historically influential person lineages) the persons in the lineages are annotated with up to five sources. The sources can be categorized by physical type (book, chart, CD, periodical, and web site) and also by structure (Complied genealogies, Genealogy lineages, History books, Published family genealogies, Genealogy publication web sites, Encyclopaedia, and Oral tradition). A specific lineage can then be characterized by the dominant physical source type and by the dominate genealogy structure. Appendix D lists all the sources identified in Appendices B and C.

The physical type categorization of the annotated sources listed in Appendix D is:

- 104 books. These include 40 volumes from Europaische Stammtafeln, Collins's Peerage of England and Cockayne's Peerage of Great Britain.

- 70 web sites. A few of the web sites may now only be available in the Internet Archive.
- 4 charts.
- 1 periodical.
- 1 CD-Rom.

By examining the correctness property of the genealogy, a rough assessment of the trustworthiness associated with a physical type can be generated:

- Published books are most trustworthy. However, care must still be taken. I analyzed Stewart's "Royalty for Commoners", source ID 107, and sent 10 pages of corrections to the author.
- Web sites are least trustworthy. There are exceptions. Lindwall's "King Erik XIVs ancestors in 14 generations", source ID 130, linked very well with published lineages from books. Lineages from the genealogy aggregation web sites (Wikitree, Geni, Familysearch) were prone to more errors the further the lineage was traced into the past.

The genealogy structural analysis sorts the source identifiers into the following categories. Each source reference is listed in Appendix D.

Compiled Genealogies—49 sources. These track families across multiple generations.

- 2, 22, 78, 95, 108, 113, 114, 149, 153, 160,
- 181, 223, 226, 234, 240, 244, 247, 258, 259, 260,
- 270, 287, 293, 315, 324, 328, 333, 334, 340, 342,
- 343, 382, 448, 651, 714, 938, 964, 1146, 1149, 1281,
- 1318, 1589, 1641, 1660, 1664, 1680, 1683, 1689, 1709.

Authoritative Compiled Genealogies: Europaische Stammtafeln, Collins's Peerage of England and Cockayne's Peerage of Great Britain–40 sources.

- 51, 58, 135, 138, 143, 144, 146, 150, 154, 155,
- 158, 161, 162, 163, 164, 165, 166, 167, 168, 169,
- 170, 171, 172, 173, 175, 179, 184, 193, 194, 195,
- 199, 200, 204, 205, 206, 212, 216, 221, 224, 239.

Genealogy Lineages—19 sources. These track lineages to or from a notable person.

- 1, 15, 42, 46, 52, 66, 67, 73, 84, 106,
- 107, 130, 262, 307, 311, 589, 852, 860, 1393.

History books—24 sources. These include lineages, family ancestors, and descendants of progenitors.

- 3, 9, 28, 30, 31, 32, 35, 37, 59, 62,
- 69, 71, 72, 80, 81, 96, 100, 104, 132, 304,
- 414, 503, 507, 1502.

Published family genealogies—34 sources. These list family ancestors and relatives.

- 49, 63, 90, 102, 109, 112, 115, 117, 119, 129,
- 152, 203, 227, 303, 415, 501, 502, 504, 575, 656,
- 763, 1137, 1154, 1155, 1162, 1163, 1191, 1205, 1251, 1263,
- 1407, 1548, 1556, 1564.

Genealogy publication web sites—5 sources. These aggregate genealogies for web access.

- 6, 711, 724, 956, 1376.

Encyclopaedia—8 sources

- 76, 268, 493, 566, 936, 1517, 1520, 1642.

Oral tradition—1 source

- 32.

As previously mentioned, trustworthy sources were found in the lists of Compiled Genealogies, Genealogy Lineages, History Books, and Encyclopaedia. The published family genealogies had to be carefully appraised. These genealogies tended to be more accurate if they traced the descendants from a progenitor.

In Appendices B and C, an analysis of each lineage is provided based upon the major sources contributing to the lineage. In summary, most lineages used information from: the following sources:

- Europaische Stammtafeln, source IDs 135, 138, 150, 154, 193, 194, 195, 195, 199, 200, 204, 205, 206, 212, 216, 221, 224, 239,
- Collins's Peerage of England, source IDs 51, 144, 146, 155, 158, 161, 175, 179, 184,
- Cockayn's Peerage of Great Britain, source IDs 58, 162, 163, 164, 165, 166, 167, 168, 169, 170, 171, 172, 173,
- Debrett's Kings and Queens of Europe, source ID 108,
- McNaughton's "The Book of Kings", source ID 78,

- Morby's "Dynasties of the World", source ID 106.

Lineages that reference these sources for each person in the lineage can be considered trustworthy.

Some lineages required connections through Genealogy publication web sites:

- Findagrave, source ID 664.
- Geni, source ID 711.
- Familypedia, source ID 724.
- Wikitree, source ID 956.
- Familysearch, source ID 6, 1376.
- MyHeritage, source ID 1472.

These lineages are trustworthy if there are additional sources that provide similar information.

Finally, some of the lineages relied upon Oral tradition:

- Snorri's "Heimskringla, History of the Kings of Norway", source ID 32 (Sturluson, 1964)

The major sources of information were Compiled Genealogies, published as books. History books were primarily used to identify specific historical persons and their immediate familial relationships. In some cases, Genealogy publication web sites had to be used to create a link to an historical person. The resulting lineages can be considered less reliable.

Reference

Sturluson, S. (1964). *Heimskringla, history of the kings of Norway*. University of Texas Press.

Summary

We parse information and knowledge relationships from multiple communications to build a personal Knowledge base about historical events. We can evaluate our knowledge base by comparing our interpretation of historical events with external Knowledge bases. If our interpretation corresponds to a community-consensus knowledge base, we can reliably communicate with members of the consensus community.

The trustworthiness of communications can be assessed by extracting their information content into a collection and annotating each element of information with a list of sources. Each collection can be analyzed for the epistemological properties of Consistency, Correctness, Closure, Connectivity, Completeness, and Coherence. The analyses establish a knowledge base that characterizes the relationships between information elements present in the collection. The analyses may be used to establish trustworthiness if relationships are then established between the collection and external community-consensus knowledge bases.

For communication with the future, trustworthiness requires the identification and preservation of the knowledge base that defines how the information content will be interpreted.

For communication from the past, trustworthiness requires the construction of a knowledge base that defines relationships between information elements, and the linking of these relationships to a current community-consensus knowledge base. Since genealogies are constructed by parsing communication from the past, the epistemological properties of the information in the genealogy are evaluated to identify trustworthy genealogies. Coherence requires an assessment of the information content to identify whether information is missing that is needed to draw conclusions about the topic of discussion. Missing information can be identified by comparing with the external community consensus database and through modeling of the topic of discussion. A property is needed that enables the evaluation of information elements essential for validating the unifying concepts within the communications and that determines whether needed information is missing.

A 330,610-person Research Genealogy was used as the starting point for constructing a genealogy knowledge base and applied to the generation of a Genealogical History of the Modern World. The properties that were explored included:

- Consistency—identification of normalized name spaces for persons, dates, locations, and titles. This included estimating a birth year for every person and identifying a country for birth locations.
- Correctness—validation that all age ranges for age at birth of a child or age at marriage or age at death fell within biologically feasible ranges.
- Closure—verification that connection paths could be found between all persons within the genealogy.
- Connectivity—identification of groupings among persons, including common ancestors of a group of persons, global connectivity (largest number of ascents over the smallest number of generations), local connectivity (largest number of times the person is the lineage coalescence point for own-cousin relationships over the smallest number of generations), ancestral distribution fraction, and descendant distribution fraction.
- Completeness—identification of a Unifying Ancestry to which all persons of Western European descent should be able to link their ancestry. The best Unifying Ancestry was based on the common ancestors of the Royal Families of Western Europe. The Unifying Ancestry was then reduced to a set of Progenitors for persons of Western European descent.
- Coherence—linkage of the Unifying ancestry to an external knowledge base consisting of Palmer's "History of the Modern World". By establishing a link, familial relationships could be found between historically important persons and members of the Research Genealogy. This establishes a Genealogical History of the Modern World that defines how Western Europeans are related to notable persons. The expectation is that as cultural groups intermarry, a common heritage emerges, unifying links to cultural groups. A Coherence metric was derived that analyzes whether essential information is missing that is needed to support the identification of progenitors of the Unifying Ancestry. The metric calculates the number of generations that must be traced to identify a progenitor, based on the lineage coalescence present within the genealogy.

The Research Genealogy was used to explore the symmetry properties of genealogies that are driven by the exponential growth of the number of ancestors per generation and the exponential growth of linage coalescence. The analyses that were used for lineage coalescence are listed in Appendix E. The analyses can be applied to any Gedcom file through the CoreGEN3 genealogy analysis workbench. The identification of lineages to Progenitors for Western Europeans validates the prediction of population genetics, that within 55 generations Progenitors can be found that are ancestors of all persons of Western European descent alive today.

The need to use multiple communications from the past (sources) is driven by the paucity of information about historical lineages. The analyzed lineages were based on the ancestry of Noble Houses and Royal Families. The sources track intermarriage between Noble Houses, leading to the identification of the degree of lineage coalescences needed to identify progenitors. The information provided by each source had to be evaluated for inclusion in the knowledge base (Unifying Ancestry). The Unifying Ancestry then serves as a proxy for lineage coalescence. Instead of tracing a lineage to a Progenitor, the lineage is only traced to the first connection with the Unifying Ancestry. The Unifying Ancestry then provides the link to the Progenitor. This makes it possible to build a Genealogical History of the World, identifying familial relationships between persons involved in historical events. The Unifying Ancestries were evaluated by analyzing the number of generations that lineages must be traced for the average number of ascents to equal the population size of the National Community. Unifying Ancestries that required fewer generations were preferentially chosen.

A major finding of the research is the existence of a Unifying Ancestry for a national community. This has multiple implications:

- For preservation, the persons referenced in historical events can be linked to the Unifying Ancestry. This provides an index across the archive, demonstrating familial relationships between persons documented by the records. A Unifying Ancestry is an essential information base for indexing preservation records.
- For history, a Genealogical History of the Modern World can be constructed, identifying relationships between Western Europeans involved in historical events.
- For genealogy, completeness can now be defined as a lineage that connects to the Unifying Ancestry. Relationships can then be found to every other person that is able to link their ancestry to the Unifying Ancestry.

Appendix A: Cousin Relationship Terminology

The cousin relationship is a function of three parameters:

1. n—the number of generations from the root person to the closest common ancestor
2. m—the number of generations from the relative to the closest common ancestor
3. h—is set if the relative's lineage terminates with a different spouse of the common ancestor

These parameters are turned into the degree of the cousin and the number of removes as follows:

If n > m.

The cousin degree is m − 1 and uses the symbol "C".
The number of removes is n − m and uses the symbol "R".

If n < m.

The cousin degree is n − 1 and uses the symbol "C".
The number of removes is m − n and uses the symbol "R*".

If the lineages link to different spouses of the closest common ancestor, the cousin relationship becomes a "half" cousin and is prepended with the letter "h"

The cousin relationships can be inverted to find the number of generations given the cousin degree, "C", and number of removes, "R".

n, the number of generations from the root person to the closest common ancestor, is C + R + 1.

© The Editor(s) (if applicable) and The Author(s), under exclusive license to Springer Nature Switzerland AG 2023
R. W. Moore, *Trustworthy Communications and Complete Genealogies*, Synthesis Lectures on Information Concepts, Retrieval, and Services,
https://doi.org/10.1007/978-3-031-16836-9

m, the number of generations from the relative to the closest common ancestor, is C + 1.

For cousin relationships defined using "R*",

n, the number of generations from the root person to the closest common ancestor, is C + 1.
m, the number of generations from the relative to the closest common ancestor, is C + R* + 1.

Note that comparing 4C2R and 4C2R*:

- If the relationship of the cousin to the root person is 4C2R,
- Then the relationship of the root person to the cousin is 4C2R*

4C2R has

$$n = 7$$
$$m = 5$$

4C2R* has

$$n = 5$$
$$m = 7$$

In Table 20, the notation h4C2R* means half 4th cousin, two removes. The root person's lineage has 5 generations to the common ancestor, while the relative's lineage has 7 generations to the common ancestor and links to a different spouse.

Appendix B: Sample Lineages to the Unifying Ancestry Progenitors

Example lineages are provided to each of the Progenitors listed in Table 3.14. Lineages of maximal ascent from Prince George of Cambridge to each Progenitor are listed. Each generation in the lineage is annotated with up to five source identifiers. Appendix D provides an ordered list of all the referenced source identifiers. In each table:

- Gen is the generation number
- Relation is the familial relationship from the previous person in the list.
- Up to 5 sources are listed for each person

For each lineage, dominant sources are listed. The abbreviations stand for:

- Web-lineage—genealogy lineages web site
- Collins—Collins's Peerage of England
- Stammtafeln—Schwennicke's Europaische Stammtaflen
- McNaughton—McNaughton's Book of Kings
- Geni—Geni.com genealogy web publication site
- Saga—Heimskringla Saga.

The lineages for each Progenitor in the following tables can be traced through the listed dominant sources:

- Potitus Photaighe of Ireland (Table B.1)

  - Web-lineage, Collins, Stammtafeln, McNaughton
  - Potitus Photaighe of Ireland is the 44th Great-Grandfather of Prince George and has 117 billion ascents from Prince George. The lineage goes through Ireland, France, Germany, Hungary, and Austria before ending in England. The lineage to Potitus was compiled by Claude Chatelain, source 1689. The lineage connects to the lineage of Maximal Ascent to Charlemagne with Lothair II of Lorraine (835–869).

R. W. Moore, *Trustworthy Communications and Complete Genealogies*, Synthesis Lectures on Information Concepts, Retrieval, and Services, https://doi.org/10.1007/978-3-031-16836-9

- Eudes of the Gewissi (Table B.2)

  - Web-lineage, Stammtafeln, McNaughton
  - Eudes of the Gewissi is of Goth descent but lived in England. He is the 44th Great-Grandfather of Prince George and has 24.6 billion ascents from Prince George. The lineage goes through Wales, France, and Germany before ending in England. The lineage connects to the lineage of Maximal Ascent to Charlemagne with Philip Mountbatten (1921–2021).

- Magnus Clemens Maximus (Table B.3)

  - Web-lineage, Stammtafeln, McNaughton
  - Magnus Clemens Maximus is of Italian descent. He is the 43rd Great-Grandfather of Prince George and has 15.8 billion ascents from Prince George. The lineage goes through Wales, France, and Germany before ending in England. The lineage connects to the lineage of Maximal Ascent to Charlemagne with Philip Mountbatten (1921–2021).

- Coel Hen (Table B.4)

  - Web-lineage, Stammtafeln, McNaughton
  - Coel Hen is the 43rd Great-Grandfather of Prince George and has 8.38 billion ascents from Prince George. The lineage goes through Wales, France, and Germany before ending in England. The lineage connects to the lineage of Maximal Ascent to Charlemagne with Philip Mountbatten (1921–2021).

- Danp of Denmark (Table B.5)

  - Saga, Collins, Stammtafeln, McNaughton
  - King Danp of Denmark is the 45th Great-Grandfather of Prince George and has 12.8 billion ascents from Prince George. The lineage goes through Denmark, Norway, Sweden, France, Hungary, Austria, and Germany before ending in England. The lineage down to France comes from the Heimskringla Saga. The lineage connects to the lineage of Maximal Ascent to Charlemagne with Gerberga of Upper Burgundy (965–1017).

- Vortigern of Powys (Table B.6)

  - Web-lineage, Geni, Stammtafeln, McNaughton
  - Vortigern of Powys is the 42nd Great-Grandfather of Prince George of Cambridge and has 9.4 billion ascents from Prince George. The lineage goes through Wales, Northern England, France, and Germany before ending in England. The lineage connects to the lineage of Maximal Ascent for Charlemagne with Philip Mountbatten (1921–2021).

- Pompeius of Dyrrhachium (Table B.7)

- Web-lineage, Stammtafeln, McNaughton
- Pompeius of Dyrrhachium was born in the Roman Empire in Albania. He is the 43rd Great-Grandfather of Prince George of Cambridge and has 15.2 billion ascents from Prince George. The lineage goes through Italy, Spain, France, and Germany before ending in England. The lineage connects to the lineage of Maximal Ascent for Charlemagne with Philip Mountbatten (1921–2021).

- Cunedda the Great (Table B.8)

  - Web-lineage, Stammtafeln, McNaughton
  - Cunedda the Great was born in Wales. He is the 42nd Great-Grandfather of Prince George of Cambridge and has 4.4 billion ascents from Prince George. The lineage goes through Wales, France, and Germany before ending in England. The lineage connects to the lineage of Maximal Ascent for Charlemagne with Philip Mountbatten (1921–2021).

- Clodgar von Theouranne (Table B.9)

  - Web-lineage, Stammtafeln, McNaughton
  - Clodgar von Therouanne is the 39th Great-Grandfather of Prince George of Cambridge and has 58.7 billion ascents from Prince George. The lineage goes through France, Germany, Hungary, and Austria before ending in England. The lineage connects to the lineage of Maximal Ascent for Charlemagne with Lothaire II of Lorraine (835–869).

- Charibert II of Neustria (Table B.10)

  - Web-lineage, Stammtafeln, McNaughton
  - Charibert II of Neustria is the 37th Great-Grandfather of Prince George and has 52.9 billion ascents from Prince George. The lineage goes through France, Germany, Hungary, and Austria before ending in England. The lineage connects to the lineage of Maximal Ascent for Charlemagne with Philip Mountbatten (1921–2021).

- Charlemagne (Table B.11)

  - Stammtafeln, McNaughton
  - Charlemagne is the 32nd Great-Grandfather of Prince George of Cambridge and has 3.24 billion ascents from Prince George. The lineage goes through France, Germany, Hungary, and Austria before ending in England. Charlemagne was the Holy Roman Emperor (800–814).

Lineages that provide references to Collins, Stammtafeln and McNaughton can be considered trustworthy. Most require additional information from lineages published on the web. Lineages to Charlemagne remain the most reliable.

Table B.1 Lineage of maximal ascent to Potitus Photaighe of Ireland from George Alexander Louis Mountbatten-Windsor

Gen	Relation	(Birth–Death)	Person	(Sources)
0	Start	(300–)	Potitus Photaighe of Ireland	(1689)
1	Son	(325–)	Calpurnius of Ireland	(1689)
2	Daughter	(355–)	Trigidia Agris of Ireland	(1689)
3	Son	(395–434)	Salomon I Cynyor de Cornouaille	(1689)
4	Son	(420–464)	Aldrien of Brittany	(1689, 1548)
5	Daughter	(462–)	Gania von Cornouaille	(1154, 1689, 1548)
6	Daughter	(510–)	Mathilde von Boulogne	(1154, 1548)
7	Son	(535–570)	Richard de Ponthieu	(1154, 1548)
8	Son	(560–614)	Ricmar d'Artois	(1683, 1564)
9	Son	(575–641)	Aega de Hesbaye	(51, 1281, 1689, 22, 328)
10	Daughter	(600–678)	Siagree of France	(107, 328, 22, 1689)
11	Daughter	(630–670)	Hymnegilde de Treves	(52, 15, 1, 1517, 1689)
12	Daughter	(647–)	Berswinde of Autun	(107, 227, 1689, 22)
13	Son	(672–722)	Adalbert de Alsace	(107, 328, 1689, 22)
14	Son	(704–767)	Luitfried I Hugues d'Alsace	(1689, 1564)
15	Son	(733–779)	Girard I of Paris	(107, 227, 328, 343, 711)
16	Daughter	(779–839)	Aba de Morvois	(216, 193, 115, 107, 303)
17	Daughter	(800–851)	Ermengarde of Alsace	(150, 193, 30, 31, 115)
18	Son	(835–869)	Lothair II of Lorraine	(51, 150, 216, 30, 107)
19	Daughter	(852–907)	Gisela of Lorraine	(150, 216, 107, 328, 1689)
20	Daughter	(868–917)	Ragnhildis von Friesland	(150, 216, 107)
21	Daughter	(896–968)	Matilda von Ringleheim	(51, 150, 2, 115, 107)
22	Daughter	(913–984)	Gerberge of Saxony	(51, 144, 150, 216, 193)
23	Daughter	(943–981)	Mathilde of France	(150, 216, 2, 30, 107)
24	Daughter	(965–1017)	Gerberga of Upper Burgundy	(150, 194, 115, 107, 181)
25	Daughter	(990–1043)	Gisela of Swabia	(150, 115, 107, 293, 311)

(continued)

Table B.1 (continued)

Gen	Relation	(Birth–Death)	Person	(Sources)
26	Son	(1017–1056)	Henry III the Mild	(216, 150, 2, 106, 114)
27	Son	(1050–1106)	Henry IV of Holy Roman Empire	(150, 216, 2, 35, 96)
28	Daughter	(1073–1143)	Agnes of Franconia	(150, 2, 35, 106, 115)
29	Daughter	(1088–1110)	Hedwig-Eilike Hohenstaufen	(150, 244, 1376)
30	Daughter	(1103–1170)	Heilika von Lengenfeld-Hopfenohe	(150, 130, 244, 328, 566)
31	Daughter	(1130–1174)	Hedwig von Wittelsbach	(150, 244)
32	Son	(1159–1204)	Berthold VI of Meran	(216, 150, 115, 107, 130)
33	Daughter	(1181–1213)	Gertrude von Meran	(150, 216, 2, 95, 113)
34	Son	(1206–1270)	Bela IV of Hungary	(150, 216, 2, 62, 95)
35	Son	(1239–1272)	Stephen V of Hungary	(216, 2, 106, 113, 115)
36	Daughter	(1257–1323)	Marie Arpad of Hungary	(216, 2, 100, 106, 113)
37	Daughter	(1273–1299)	Marguerite de Naples	(216, 2, 69, 84, 90)
38	Daughter	(1293–1342)	Jeanne of Valois	(216, 2, 95, 115, 107)
39	Daughter	(1310–1356)	Marguerite of Hainault	(216, 150, 106, 260, 1393)
40	Son	(1336–1404)	Albert I of Bavaria	(150, 216, 224, 239, 206)
41	Daughter	(1377–1410)	Joanna of Bavaria	(150, 1393, 63)
42	Son	(1397–1439)	Albert II of Austria	(150, 216, 224, 63, 106)
43	Daughter	(1436–1505)	Elizabeth of Austria	(150, 216, 113, 119, 258)
44	Daughter	(1464–1512)	Sofia of Poland	(216, 150, 258, 589)
45	Son	(1481–1527)	Kasimir of Brandenburg-Bayreuth	(193, 150, 589)
46	Daughter	(1519–1567)	Maria of Brandenburg-Bayreuth	(150, 589)
47	Daughter	(1544–1592)	Dorothea Susanna of Simmern	(150, 311, 589)
48	Son	(1570–1605)	Johann III of Saxe-Weimar	(150, 193, 78, 311, 589)
49	Son	(1601–1675)	Ernst I of Saxe-Gotha	(193, 150, 78, 258, 311)
50	Son	(1646–1691)	Friedrich I of Saxe-Gotha	(150, 194, 193, 78, 311)
51	Daughter	(1670–1728)	Anna Sophia of Saxe-Gotha	(150, 194, 852)
52	Daughter	(1700–1780)	Anna Sophia von Schwarzburg-Rudolstadt	(78, 194, 150, 589, 852)

(continued)

Table B.1 (continued)

Gen	Relation	(Birth–Death)	Person	(Sources)
53	Daughter	(1731–1810)	Charlotte Sophia of Saxe-Coburg-Saalfeld	(150, 194, 78, 589, 852)
54	Son	(1756–1837)	Friedrich Franz of Mecklenburg-Schwerin	(150, 194, 78, 311, 589)
55	Daughter	(1779–1801)	Luise Charlotte of Mecklenburg-Schwerin	(150, 194, 78, 311, 589)
56	Daughter	(1800–1831)	Dorothy Luise of Saxe-Gotha-Altenburg	(78, 150, 311, 589, 852)
57	Son	(1819–1861)	Francis Albert of Saxe-Coburg-Gotha	(78, 150, 216, 193, 194)
58	Daughter	(1843–1878)	Alice Maud Mary of Saxe-Coburg-Gotha	(78, 216, 193, 224, 46)
59	Daughter	(1863–1950)	Victoria of Hesse and the Rhine	(78, 193, 224, 239, 46)
60	Daughter	(1885–1969)	Victoria Alice Elizabeth of Battenberg	(78, 216, 224, 239, 46)
61	Son	(1921–2021)	Philip Mountbatten	(78, 216, 224, 46, 52)
62	Son	(1948–)	Charles Philip Mountbatten-Windsor	(78, 216, 224, 46, 73)
63	Son	(1982–)	William Arthur Mountbatten-Windsor	(216, 95, 112, 113, 223)
64	Son	(2013–)	George Alexander Mountbatten-Windsor	(493, 1318)

Table B.2 Lineage of maximal ascent to Eudes of the Gewissi from George Alexander Louis Mountbatten-Windsor

(Birth–Death)	Relation		Person	(Sources)
0	Start	(320–)	Eudes of the Gewissi	(1, 52, 95, 1251, 711)
1	Son	(345–)	Cynan I ap Eudaf Hen	(1, 95, 303, 1251, 711)
2	Son	(364–)	Cadfan of the Gewissi	(95)
3	Daughter	(384–)	Ystrafael of Britain	(1, 52, 95, 328)
4	Daughter	(409–)	Gwawl of Gwynedd	(1, 52, 95, 328, 493)
5	Son	(442–)	Ceredig of Gwynedd	(1, 52, 73, 95, 328)
6	Son	(460–)	Corwyn of Gwynedd	(1, 52, 95, 328)
7	Daughter	(481–)	D'Anaumide verch Teithfalit	(303, 328, 95)

(continued)

Table B.2 (continued)

(Birth–Death)	Relation		Person	(Sources)
8	Daughter	(530–)	Alienor of the Bretons	(1)
9	Son	(556–)	Withur de Leon d'Acqs	(964)
10	Son	(572–)	Ausoch de Leon d'Acqs	(1, 303, 493)
11	Daughter	(587–)	Pritelle de Leon d'Acqs	(1, 328, 493, 1689)
12	Son	(602–658)	Judicael II de Cornouaille	(1, 95, 107, 303, 328)
13	Son	(632–711)	Gradlon Flam de Cornouaille	(956, 1689, 711)
14	Daughter	(660–)	Cheronnoge Congare de Cornouaille	(1689)
15	Son	(680–720)	Riwallon II de Poher de Bors	(1154, 711)
16	Daughter	(720–)	Marmoec de Poher	(1154, 711)
17	Son	(765–812)	Erispoe I Ernoch de Poher	(1641, 1689, 1154)
18	Son	(790–851)	Nominoe de Vannes	(1, 52, 95, 107, 328)
19	Son	(815–857)	Erispoe II de Vannes	(1, 52, 95, 107, 1689)
20	Daughter	(830–880)	Lotitia de Poher	(328, 1154, 1689, 107)
21	Son	(860–888)	Berenger Judicael I of Rennes	(216, 328, 1689, 107)
22	Daughter	(883–925)	Papie de Bayeux	(216, 2, 95, 115, 107)
23	Daughter	(917–969)	Gerloc of Normandy	(216, 2, 95, 115, 107)
24	Daughter	(945–1006)	Adelais of Aquitaine	(216, 2, 30, 84, 95)
25	Son	(972–1031)	Robert II the Pious	(175, 216, 2, 30, 52)
26	Daughter	(1009–1079)	Adelaide of France	(216, 2, 35, 52, 114)
27	Son	(1035–1093)	Robert I of Flanders	(150, 216, 193, 35, 106)
28	Daughter	(1070–1117)	Gertrude of Flanders	(216, 193, 106, 114, 115)
29	Daughter	(1087–)	Gisela of Lorraine	(193, 1709, 107, 382)
30	Daughter	(1115–1147)	Agnes von Saarbrucken	(150, 382, 130)
31	Daughter	(1135–1191)	Jutta of Swabia	(150, 130)
32	Son	(1155–1217)	Hermann I of Thuringia	(150, 193, 224, 130)
33	Daughter	(1198–1244)	Irmgard of Thuringia	(193, 150, 130)

(continued)

Table B.2 (continued)

(Birth–Death)	Relation		Person	(Sources)
34	Daughter	(1215–1277)	Jutta of Anhalt	(193, 194, 130)
35	Son	(1235–1291)	Heinrich I of Werle-Gustrow	(51, 150, 216, 194, 130)
36	Daughter	(1268–1312)	Richiza of Werle-Gustrow	(51, 150, 194, 1393, 130)
37	Son	(1304–1369)	Magnus I of Brunswick	(51, 150, 205, 193, 224)
38	Son	(1328–1373)	Magnus II of Brunswick	(51, 150, 194, 193, 224)
39	Son	(1355–1416)	Henry II of Brunswick-Luneburg	(51, 150, 193, 130, 311)
40	Daughter	(1388–1442)	Catharina of Brunswick-Luneburg	(150, 130, 311, 1393)
41	Son	(1412–1464)	Friedrich II of Saxony	(150, 78, 106, 119, 311)
42	Son	(1441–1486)	Ernst I of Saxony	(51, 150, 194, 78, 106)
43	Daughter	(1461–1521)	Christine of Saxony	(150, 194, 113, 119, 589)
44	Daughter	(1485–1555)	Elizabeth of Denmark	(150, 194, 119, 258, 1393)
45	Daughter	(1511–1577)	Margarete of Brandenburg	(193, 150, 129, 258, 589)
46	Son	(1536–1586)	Joachim Ernst II of Anhalt-Zerbst	(150, 193, 194, 78, 117)
47	Daughter	(1563–1607)	Elisabeth of Anhalt-Zerbst	(150, 193, 589, 119, 566)
48	Daughter	(1582–1616)	Magdalena von Brandenburg	(150, 193, 589, 566)
49	Daughter	(1601–1659)	Anna Eleonore of Hesse-Darmstadt	(51, 150, 193, 258, 589)
50	Son	(1629–1698)	Ernst Ausgustus of Brunswick-Luneburg	(78, 51, 150, 163, 52)
51	Son	(1660–1727)	George I Louis of Hanover	(78, 51, 150, 170, 163)
52	Son	(1683–1760)	George II August of Hanover	(78, 51, 144, 150, 193)
53	Daughter	(1723–1772)	Mary of Hanover	(78, 51, 144, 150, 193)
54	Son	(1747–1837)	Frederick III of Hesse-Cassel	(78, 144, 150, 193, 194)
55	Son	(1787–1867)	William of Hesse-Rumpenheim	(78, 193, 216, 194, 224)

(continued)

Table B.2 (continued)

(Birth–Death)	Relation		Person	(Sources)
56	Daughter	(1817–1898)	Louise Wilhelmina of Hesse-Cassel	(78, 216, 193, 194, 224)
57	Son	(1845–1913)	George I of the Hellenes	(78, 216, 194, 224, 46)
58	Son	(1882–1944)	Andrew of Greece and Denmark	(78, 150, 216, 193, 224)
59	Son	(1921–2021)	Philip Mountbatten	(78, 216, 224, 46, 52)
60	Son	(1948–)	Charles Philip Mountbatten-Windsor	(78, 216, 224, 46, 73)
61	Son	(1982–)	William Arthur Mountbatten-Windsor	(216, 95, 112, 113, 223)
62	Son	(2013–)	George Alexander Mountbatten-Windsor	(493, 1318)

Table B.3 Lineage of maximal ascent to Magnus Clemens Maximus from George Alexander Louis Mountbatten-Windsor

Gen	Relation	(Birth–Death)	Person	(Sources)
0	Start	(340–388)	Magnus Clemens Maximus	(3, 1, 9, 28, 52)
1	Daughter	(387–)	Severa of Britain	(1, 9, 28, 52, 95)
2	Daughter	(424–)	Ribrawst of Powys	(1, 52, 1251)
3	Daughter	(442–)	Meleri of Breichniog	(1, 52, 1251)
4	Son	(460–)	Corwyn of Gwynedd	(1, 52, 95, 328)
5	Daughter	(481–)	D'Anaumide verch Teithfalit	(303, 328, 95)
6	Daughter	(530–)	Alienor of the Bretons	(1)
7	Son	(556–)	Withur de Leon d'Acqs	(964)
8	Son	(572–)	Ausoch de Leon d'Acqs	(1, 303, 493)
9	Daughter	(587–)	Pritelle de Leon d'Acqs	(1, 328, 493, 1689)
10	Son	(602–658)	Judicael II de Cornouaille	(1, 95, 107, 303, 328)
11	Son	(632–711)	Gradlon Flam de Cornouaille	(956, 1689, 711)
12	Daughter	(660–)	Cheronnoge Congare de Cornouaille	(1689)

(continued)

Table B.3 (continued)

Gen	Relation	(Birth–Death)	Person	(Sources)
13	Son	(680–720)	Riwallon II de Poher de Bors	(1154, 711)
14	Daughter	(720–)	Marmoec de Poher	(1154, 711)
15	Son	(765–812)	Erispoe I Ernoch de Poher	(1641, 1689, 1154)
16	Son	(790–851)	Nominoe de Vannes	(1, 52, 95, 107, 328)
17	Son	(815–857)	Erispoe II de Vannes	(1, 52, 95, 107, 1689)
18	Daughter	(830–880)	Lotitia de Poher	(328, 1154, 1689, 107)
19	Son	(860–888)	Berenger Judicael I of Rennes	(216, 328, 1689, 107)
20	Daughter	(883–925)	Papie de Bayeux	(216, 2, 95, 115, 107)
21	Daughter	(917–969)	Gerloc of Normandy	(216, 2, 95, 115, 107)
22	Daughter	(945–1006)	Adelais of Aquitaine	(216, 2, 30, 84, 95)
23	Son	(972–1031)	Robert II the Pious	(175, 216, 2, 30, 52)
24	Daughter	(1009–1079)	Adelaide of France	(216, 2, 35, 52, 114)
25	Son	(1035–1093)	Robert I of Flanders	(150, 216, 193, 35, 106)
26	Daughter	(1070–1117)	Gertrude of Flanders	(216, 193, 106, 114, 115)
27	Daughter	(1087–)	Gisela of Lorraine	(193, 1709, 107, 382)
28	Daughter	(1115–1147)	Agnes von Saarbrucken	(150, 382, 130)
29	Daughter	(1135–1191)	Jutta of Swabia	(150, 130)
30	Son	(1155–1217)	Hermann I of Thuringia	(150, 193, 224, 130)
31	Daughter	(1198–1244)	Irmgard of Thuringia	(193, 150, 130)
32	Daughter	(1215–1277)	Jutta of Anhalt	(193, 194, 130)
33	Son	(1235–1291)	Heinrich I of Werle-Gustrow	(51, 150, 216, 194, 130)
34	Daughter	(1268–1312)	Richiza of Werle-Gustrow	(51, 150, 194, 1393, 130)
35	Son	(1304–1369)	Magnus I of Brunswick	(51, 150, 205, 193, 224)
36	Son	(1328–1373)	Magnus II of Brunswick	(51, 150, 194, 193, 224)
37	Son	(1355–1416)	Henry II of Brunswick-Luneburg	(51, 150, 193, 130, 311)
38	Daughter	(1388–1442)	Catharina of Brunswick-Luneburg	(150, 130, 311, 1393)
39	Son	(1412–1464)	Friedrich II of Saxony	(150, 78, 106, 119, 311)
40	Son	(1441–1486)	Ernst I of Saxony	(51, 150, 194, 78, 106)
41	Daughter	(1461–1521)	Christine of Saxony	(150, 194, 113, 119, 589)
42	Daughter	(1485–1555)	Elizabeth of Denmark	(150, 194, 119, 258, 1393)

(continued)

Table B.3 (continued)

Gen	Relation	(Birth–Death)	Person	(Sources)
43	Daughter	(1511–1577)	Margarete of Brandenburg	(193, 150, 129, 258, 589)
44	Son	(1536–1586)	Joachim Ernst II of Anhalt-Zerbst	(150, 193, 194, 78, 117)
45	Daughter	(1563–1607)	Elisabeth of Anhalt-Zerbst	(150, 193, 589, 119, 566)
46	Daughter	(1582–1616)	Magdalena von Brandenburg	(150, 193, 589, 566)
47	Daughter	(1601–1659)	Anna Eleonore of Hesse-Darmstadt	(51, 150, 193, 258, 589)
48	Son	(1629–1698)	Ernst Ausgustus of Brunswick-Luneburg	(78, 51, 150, 163, 52)
49	Son	(1660–1727)	George I Louis of Hanover	(78, 51, 150, 170, 163)
50	Son	(1683–1760)	George II August of Hanover	(78, 51, 144, 150, 193)
51	Daughter	(1723–1772)	Mary of Hanover	(78, 51, 144, 150, 193)
52	Son	(1747–1837)	Frederick III of Hesse-Cassel	(78, 144, 150, 193, 194)
53	Son	(1787–1867)	William of Hesse-Rumpenheim	(78, 193, 216, 194, 224)
54	Daughter	(1817–1898)	Louise Wilhelmina of Hesse-Cassel	(78, 216, 193, 194, 224)
55	Son	(1845–1913)	George I of the Hellenes	(78, 216, 194, 224, 46)
56	Son	(1882–1944)	Andrew of Greece and Denmark	(78, 150, 216, 193, 224)
57	Son	(1921–2021)	Philip Mountbatten	(78, 216, 224, 46, 52)
58	Son	(1948–)	Charles Philip Mountbatten-Windsor	(78, 216, 224, 46, 73)
59	Son	(1982–)	William Arthur Mountbatten-Windsor	(216, 95, 112, 113, 223)
60	Son	(2013–)	George Alexander Mountbatten-Windsor	(493, 1318)

Table B.4 Lineage of maximal ascent to Coel Hen from George Alexander Louis Mountbatten-Windsor

Gen	Relation	(Birth–Death)	Person	(Sources)
0	Start	(380–430)	Coel Hen	(52, 95, 143, 328, 1251)
1	Daughter	(409–)	Gwawl of Gwynedd	(1, 52, 95, 328, 493)
2	Son	(442–)	Ceredig of Gwynedd	(1, 52, 73, 95, 328)
3	Son	(460–)	Corwyn of Gwynedd	(1, 52, 95, 328)
4	Daughter	(481–)	D'Anaumide verch Teithfalit	(303, 328, 95)
5	Daughter	(530–)	Alienor of the Bretons	(1)
6	Son	(556–)	Withur de Leon d'Acqs	(964)
7	Son	(572–)	Ausoch de Leon d'Acqs	(1, 303, 493)
8	Daughter	(587–)	Pritelle de Leon d'Acqs	(1, 328, 493, 1689)
9	Son	(602–658)	Judicael II de Cornouaille	(1, 95, 107, 303, 328)
10	Son	(632–711)	Gradlon Flam de Cornouaille	(956, 1689, 711)
11	Daughter	(660–)	Cheronnoge Congare de Cornouaille	(1689)
12	Son	(680–720)	Riwallon II de Poher de Bors	(1154, 711)
13	Daughter	(720–)	Marmoec de Poher	(1154, 711)
14	Son	(765–812)	Erispoe I Ernoch de Poher	(1641, 1689, 1154)
15	Son	(790–851)	Nominoe de Vannes	(1, 52, 95, 107, 328)
16	Son	(815–857)	Erispoe II de Vannes	(1, 52, 95, 107, 1689)
17	Daughter	(830–880)	Lotitia de Poher	(328, 1154, 1689, 107)
18	Son	(860–888)	Berenger Judicael I of Rennes	(216, 328, 1689, 107)
19	Daughter	(883–925)	Papie de Bayeux	(216, 2, 95, 115, 107)
20	Daughter	(917–969)	Gerloc of Normandy	(216, 2, 95, 115, 107)
21	Daughter	(945–1006)	Adelais of Aquitaine	(216, 2, 30, 84, 95)
22	Son	(972–1031)	Robert II the Pious	(175, 216, 2, 30, 52)
23	Daughter	(1009–1079)	Adelaide of France	(216, 2, 35, 52, 114)
24	Son	(1035–1093)	Robert I of Flanders	(150, 216, 193, 35, 106)
25	Daughter	(1070–1117)	Gertrude of Flanders	(216, 193, 106, 114, 115)
26	Daughter	(1087–)	Gisela of Lorraine	(193, 1709, 107, 382)
27	Daughter	(1115–1147)	Agnes von Saarbrucken	(150, 382, 130)

(continued)

Table B.4 (continued)

Gen	Relation	(Birth–Death)	Person	(Sources)
28	Daughter	(1135–1191)	Jutta of Swabia	(150, 130)
29	Son	(1155–1217)	Hermann I of Thuringia	(150, 193, 224, 130)
30	Daughter	(1198–1244)	Irmgard of Thuringia	(193, 150, 130)
31	Daughter	(1215–1277)	Jutta of Anhalt	(193, 194, 130)
32	Son	(1235–1291)	Heinrich I of Werle-Gustrow	(51, 150, 216, 194, 130)
33	Daughter	(1268–1312)	Richiza of Werle-Gustrow	(51, 150, 194, 1393, 130)
34	Son	(1304–1369)	Magnus I of Brunswick	(51, 150, 205, 193, 224)
35	Son	(1328–1373)	Magnus II of Brunswick	(51, 150, 194, 193, 224)
36	Son	(1355–1416)	Henry II of Brunswick-Luneburg	(51, 150, 193, 130, 311)
37	Daughter	(1388–1442)	Catharina of Brunswick-Luneburg	(150, 130, 311, 1393)
38	Son	(1412–1464)	Friedrich II of Saxony	(150, 78, 106, 119, 311)
39	Son	(1441–1486)	Ernst I of Saxony	(51, 150, 194, 78, 106)
40	Daughter	(1461–1521)	Christine of Saxony	(150, 194, 113, 119, 589)
41	Daughter	(1485–1555)	Elizabeth of Denmark	(150, 194, 119, 258, 1393)
42	Daughter	(1511–1577)	Margarete of Brandenburg	(193, 150, 129, 258, 589)
43	Son	(1536–1586)	Joachim Ernst II of Anhalt-Zerbst	(150, 193, 194, 78, 117)
44	Daughter	(1563–1607)	Elisabeth of Anhalt-Zerbst	(150, 193, 589, 119, 566)
45	Daughter	(1582–1616)	Magdalena von Brandenburg	(150, 193, 589, 566)
46	Daughter	(1601–1659)	Anna Eleonore of Hesse-Darmstadt	(51, 150, 193, 258, 589)
47	Son	(1629–1698)	Ernst Ausgustus of Brunswick-Luneburg	(78, 51, 150, 163, 52)
48	Son	(1660–1727)	George I Louis of Hanover	(78, 51, 150, 170, 163)
49	Son	(1683–1760)	George II August of Hanover	(78, 51, 144, 150, 193)
50	Daughter	(1723–1772)	Mary of Hanover	(78, 51, 144, 150, 193)
51	Son	(1747–1837)	Frederick III of Hesse-Cassel	(78, 144, 150, 193, 194)

(continued)

Table B.4 (continued)

Gen	Relation	(Birth–Death)	Person	(Sources)
52	Son	(1787–1867)	William of Hesse-Rumpenheim	(78, 193, 216, 194, 224)
53	Daughter	(1817–1898)	Louise Wilhelmina of Hesse-Cassel	(78, 216, 193, 194, 224)
54	Son	(1845–1913)	George I of the Hellenes	(78, 216, 194, 224, 46)
55	Son	(1882–1944)	Andrew of Greece and Denmark	(78, 150, 216, 193, 224)
56	Son	(1921–2021)	Philip Mountbatten	(78, 216, 224, 46, 52)
57	Son	(1948–)	Charles Philip Mountbatten-Windsor	(78, 216, 224, 46, 73)
58	Son	(1982–)	William Arthur Mountbatten-Windsor	(216, 95, 112, 113, 223)
59	Son	(2013–)	George Alexander Mountbatten-Windsor	(493, 1318)

Table B.5 Lineage of maximal ascent to Danp of Denmark from George Alexander Louis Mountbatten-Windsor

Gen	Relation	(Birth–Death)	Person	(Sources)
0	Start	(384–)	Danp of Denmark	(32, 107, 1520, 493)
1	Son	(404–)	Dan the Proud	(32, 66, 95, 1520)
2	Son	(433–)	Frodi Mikillati	(32, 66, 95, 107, 493)
3	Son	(456–)	Fridleif Frodisson	(32, 66, 95, 107)
4	Son	(479–548)	Frode VII the Valiant	(66, 95, 107)
5	Son	(503–)	Halfdan Frodasson	(32, 66, 95, 107)
6	Son	(526–)	Hroar Halfdansson	(66, 95, 107, 566)
7	Son	(547–)	Valdar the Mild	(66, 95, 107, 566)
8	Son	(568–)	Harald the Old	(6, 66, 95, 107, 566)
9	Son	(590–650)	Halfdan the Valiant	(32, 6, 66, 95, 107)
10	Son	(625–667)	Trond Haroldson	(711, 566)
11	Son	(655–)	Eystein the Severe	(107, 711, 32, 566, 1520)
12	Daughter	(680–)	Aasa Eysteinsdatter	(32, 107, 566, 1520, 1689)

(continued)

Table B.5 (continued)

Gen	Relation	(Birth–Death)	Person	(Sources)
13	Son	(710–780)	Eystein I of Westfold	(32, 95, 114, 107, 328)
14	Daughter	(750–807)	Geva of Westfold	(51, 216, 107, 328, 956)
15	Daughter	(775–833)	Eigilwich of Saxony	(51, 150, 31, 107, 181)
16	Daughter	(801–843)	Judith of Bavaria	(51, 150, 30, 52, 107)
17	Daughter	(820–874)	Gisela of France	(150, 216, 2, 30, 52)
18	Daughter	(838–902)	Waldrade Judith of Friuli	(216, 107, 1205, 1689, 109)
19	Son	(870–899)	Eberhard I von Thurgau	(150, 1205, 107)
20	Daughter	(889–958)	Regilinda von Thurgau	(150, 328, 107)
21	Daughter	(907–966)	Bertha of Swabia	(216, 115, 107, 181, 328)
22	Son	(925–993)	Conrad I of Burgundy	(150, 216, 194, 2, 30)
23	Daughter	(965–1017)	Gerberga of Upper Burgundy	(150, 194, 115, 107, 181)
24	Daughter	(990–1043)	Gisela of Swabia	(150, 115, 107, 293, 311)
25	Son	(1017–1056)	Henry III the Mild	(216, 150, 2, 106, 114)
26	Son	(1050–1106)	Henry IV of Holy Roman Empire	(150, 216, 2, 35, 96)
27	Daughter	(1073–1143)	Agnes of Franconia	(150, 2, 35, 106, 115)
28	Daughter	(1088–1110)	Hedwig-Eilike Hohenstaufen	(150, 244, 1376)
29	Daughter	(1103–1170)	Heilika von Lengenfeld-Hopfenohe	(150, 130, 244, 328, 566)
30	Daughter	(1130–1174)	Hedwig von Wittelsbach	(150, 244)
31	Son	(1159–1204)	Berthold VI of Meran	(216, 150, 115, 107, 130)
32	Daughter	(1181–1213)	Gertrude von Meran	(150, 216, 2, 95, 113)
33	Son	(1206–1270)	Bela IV of Hungary	(150, 216, 2, 62, 95)
34	Son	(1239–1272)	Stephen V of Hungary	(216, 2, 106, 113, 115)
35	Daughter	(1257–1323)	Marie Arpad of Hungary	(216, 2, 100, 106, 113)
36	Daughter	(1273–1299)	Marguerite de Naples	(216, 2, 69, 84, 90)
37	Daughter	(1293–1342)	Jeanne of Valois	(216, 2, 95, 115, 107)
38	Daughter	(1310–1356)	Marguerite of Hainault	(216, 150, 106, 260, 1393)
39	Son	(1336–1404)	Albert I of Bavaria	(150, 216, 224, 239, 206)
40	Daughter	(1377–1410)	Joanna of Bavaria	(150, 1393, 63)
41	Son	(1397–1439)	Albert II of Austria	(150, 216, 224, 63, 106)
42	Daughter	(1436–1505)	Elizabeth of Austria	(150, 216, 113, 119, 258)

(continued)

Table B.5 (continued)

Gen	Relation	(Birth–Death)	Person	(Sources)
43	Daughter	(1464–1512)	Sofia of Poland	(216, 150, 258, 589)
44	Son	(1481–1527)	Kasimir of Brandenburg-Bayreuth	(193, 150, 589)
45	Daughter	(1519–1567)	Maria of Brandenburg-Bayreuth	(150, 589)
46	Daughter	(1544–1592)	Dorothea Susanna of Simmern	(150, 311, 589)
47	Son	(1570–1605)	Johann III of Saxe-Weimar	(150, 193, 78, 311, 589)
48	Son	(1601–1675)	Ernst I of Saxe-Gotha	(193, 150, 78, 258, 311)
49	Son	(1646–1691)	Friedrich I of Saxe-Gotha	(150, 194, 193, 78, 311)
50	Daughter	(1670–1728)	Anna Sophia of Saxe-Gotha	(150, 194, 852)
51	Daughter	(1700–1780)	Anna Sophia von Schwarzburg-Rudolstadt	(78, 194, 150, 589, 852)
52	Daughter	(1731–1810)	Charlotte Sophia of Saxe-Coburg-Saalfeld	(150, 194, 78, 589, 852)
53	Son	(1756–1837)	Friedrich Franz of Mecklenburg-Schwerin	(150, 194, 78, 311, 589)
54	Daughter	(1779–1801)	Luise Charlotte of Mecklenburg-Schwerin	(150, 194, 78, 311, 589)
55	Daughter	(1800–1831)	Dorothy Luise of Saxe-Gotha-Altenburg	(78, 150, 311, 589, 852)
56	Son	(1819–1861)	Francis Albert of Saxe-Coburg-Gotha	(78, 150, 216, 193, 194)
57	Daughter	(1843–1878)	Alice Maud Mary of Saxe-Coburg-Gotha	(78, 216, 193, 224, 46)
58	Daughter	(1863–1950)	Victoria of Hesse and the Rhine	(78, 193, 224, 239, 46)
59	Daughter	(1885–1969)	Victoria Alice Elizabeth of Battenberg	(78, 216, 224, 239, 46)
60	Son	(1921–2021)	Philip Mountbatten	(78, 216, 224, 46, 52)
61	Son	(1948–)	Charles Philip Mountbatten-Windsor	(78, 216, 224, 46, 73)
62	Son	(1982–)	William Arthur Mountbatten-Windsor	(216, 95, 112, 113, 223)
63	Son	(2013–)	George Alexander Mountbatten-Windsor	(493, 1318)

Table B.6. Lineage of maximal ascent to Vortigern of Powys from George Alexander Louis Mountbatten-Windsor

Gen	Relation	(Birth–Death)	Person	(Source numbers)
0	Start	(395–480)	Vortigern of Powys	(1, 9, 28, 52, 73)
1	Daughter	(424–)	Ribrawst of Powys	(1, 52, 1251)
2	Daughter	(442–)	Meleri of Breichniog	(1, 52, 1251)
3	Son	(460–)	Corwyn of Gwynedd	(1, 52, 95, 328)
4	Daughter	(481–)	D'Anaumide verch Teithfalit	(303, 328, 95)
5	Daughter	(530–)	Alienor of the Bretons	(1)
6	Son	(556–)	Withur de Leon d'Acqs	(964)
7	Son	(572–)	Ausoch de Leon d'Acqs	(1, 303, 493)
8	Daughter	(587–)	Pritelle de Leon d'Acqs	(1, 328, 493, 1689)
9	Son	(602–658)	Judicael II de Cornouaille	(1, 95, 107, 303, 328)
10	Son	(632–711)	Gradlon Flam de Cornouaille	(956, 1689, 711)
11	Daughter	(660–)	Cheronnoge Congare de Cornouaille	(1689)
12	Son	(680–720)	Riwallon II de Poher de Bors	(1154, 711)
13	Daughter	(720–)	Marmoec de Poher	(1154, 711)
14	Son	(765–812)	Erispoe I Ernoch de Poher	(1641, 1689, 1154)
15	Son	(790–851)	Nominoe de Vannes	(1, 52, 95, 107, 328)
16	Son	(815–857)	Erispoe II de Vannes	(1, 52, 95, 107, 1689)
17	Daughter	(830–880)	Lotitia de Poher	(328, 1154, 1689, 107)
18	Son	(860–888)	Berenger Judicael I of Rennes	(216, 328, 1689, 107)
19	Daughter	(883–925)	Papie de Bayeux	(216, 2, 95, 115, 107)
20	Daughter	(917–969)	Gerloc of Normandy	(216, 2, 95, 115, 107)
21	Daughter	(945–1006)	Adelais of Aquitaine	(216, 2, 30, 84, 95)
22	Son	(972–1031)	Robert II the Pious	(175, 216, 2, 30, 52)
23	Daughter	(1009–1079)	Adelaide of France	(216, 2, 35, 52, 114)
24	Son	(1035–1093)	Robert I of Flanders	(150, 216, 193, 35, 106)
25	Daughter	(1070–1117)	Gertrude of Flanders	(216, 193, 106, 114, 115)
26	Daughter	(1087–)	Gisela of Lorraine	(193, 1709, 107, 382)
27	Daughter	(1115–1147)	Agnes von Saarbrucken	(150, 382, 130)

(continued)

Table B.6. (continued)

Gen	Relation	(Birth–Death)	Person	(Source numbers)
28	Daughter	(1135–1191)	Jutta of Swabia	(150, 130)
29	Son	(1155–1217)	Hermann I of Thuringia	(150, 193, 224, 130)
30	Daughter	(1198–1244)	Irmgard of Thuringia	(193, 150, 130)
31	Daughter	(1215–1277)	Jutta of Anhalt	(193, 194, 130)
32	Son	(1235–1291)	Heinrich I of Werle-Gustrow	(51, 150, 216, 194, 130)
33	Daughter	(1268–1312)	Richiza of Werle-Gustrow	(51, 150, 194, 1393, 130)
34	Son	(1304–1369)	Magnus I of Brunswick	(51, 150, 205, 193, 224)
35	Son	(1328–1373)	Magnus II of Brunswick	(51, 150, 194, 193, 224)
36	Son	(1355–1416)	Henry II of Brunswick-Luneburg	(51, 150, 193, 130, 311)
37	Daughter	(1388–1442)	Catharina of Brunswick-Luneburg	(150, 130, 311, 1393)
38	Son	(1412–1464)	Friedrich II of Saxony	(150, 78, 106, 119, 311)
39	Son	(1441–1486)	Ernst I of Saxony	(51, 150, 194, 78, 106)
40	Daughter	(1461–1521)	Christine of Saxony	(150, 194, 113, 119, 589)
41	Daughter	(1485–1555)	Elizabeth of Denmark	(150, 194, 119, 258, 1393)
42	Daughter	(1511–1577)	Margarete of Brandenburg	(193, 150, 129, 258, 589)
43	Son	(1536–1586)	Joachim Ernst II of Anhalt-Zerbst	(150, 193, 194, 78, 117)
44	Daughter	(1563–1607)	Elisabeth of Anhalt-Zerbst	(150, 193, 589, 119, 566)
45	Daughter	(1582–1616)	Magdalena von Brandenburg	(150, 193, 589, 566)
46	Daughter	(1601–1659)	Anna Eleonore of Hesse-Darmstadt	(51, 150, 193, 258, 589)
47	Son	(1629–1698)	Ernst Ausgustus of Brunswick-Luneburg	(78, 51, 150, 163, 52)
48	Son	(1660–1727)	George I Louis of Hanover	(78, 51, 150, 170, 163)
49	Son	(1683–1760)	George II August of Hanover	(78, 51, 144, 150, 193)
50	Daughter	(1723–1772)	Mary of Hanover	(78, 51, 144, 150, 193)
51	Son	(1747–1837)	Frederick III of Hesse-Cassel	(78, 144, 150, 193, 194)

(continued)

Table B.6. (continued)

Gen	Relation	(Birth–Death)	Person	(Source numbers)
52	Son	(1787–1867)	William of Hesse-Rumpenheim	(78, 193, 216, 194, 224)
53	Daughter	(1817–1898)	Louise Wilhelmina of Hesse-Cassel	(78, 216, 193, 194, 224)
54	Son	(1845–1913)	George I of the Hellenes	(78, 216, 194, 224, 46)
55	Son	(1882–1944)	Andrew of Greece and Denmark	(78, 150, 216, 193, 224)
56	Son	(1921–2021)	Philip Mountbatten	(78, 216, 224, 46, 52)
57	Son	(1948–)	Charles Philip Mountbatten-Windsor	(78, 216, 224, 46, 73)
58	Son	(1982–)	William Arthur Mountbatten-Windsor	(216, 95, 112, 113, 223)
59	Son	(2013–)	George Alexander Mountbatten-Windsor	(493, 1318)

Table B.7 Lineage of maximal ascent to Pompeius of Dyrrhachium from George Alexander Louis Mountbatten-Windsor

Gen	Relation	(Birth–Death)	Person	(Source numbers)
0	Start	(400–)	Pompeius of Dyrrhachium	(860, 711)
1	Son	(440–496)	Paulus of Eastern Rome	(860)
2	Son	(478–518)	Probus Magnus	(860)
3	Son	(498–525)	Probus of Italy	(860)
4	Daughter	(522–)	Juliana of Eastern Rome	(3, 860)
5	Daughter	(544–)	Anastasia Areobinda	(860)
6	Daughter	(567–)	Flavia Juliana	(216, 2, 860, 1689)
7	Son	(605–)	Ardabusto of Spain	(216, 2, 860, 1517, 1642)
8	Son	(632–687)	Ervigo of Visigoths	(216, 2, 37, 107, 860)
9	Son	(680–750)	Pedro I of Cantabria	(216, 2, 81, 107, 860)
10	Daughter	(700–)	Wanda of the Visigoths	(1689)

(continued)

Table B.7 (continued)

Gen	Relation	(Birth–Death)	Person	(Source numbers)
11	Daughter	(720–)	Adele of Gascogne	(2, 1689)
12	Son	(740–778)	Sancho Loup II of Gascogne	(2, 1641, 1689)
13	Son	(770–816)	Sanche I Loup de Gascogne	(2, 1689, 107, 1641)
14	Daughter	(804–843)	Dhuoda of Gascogne	(31, 52, 107, 227, 315)
15	Son	(841–886)	Bernard III of Aquitaine	(216, 1, 31, 52, 106)
16	Daughter	(860–906)	Adelinde of Aquitaine	(216, 106, 1689, 30)
17	Daughter	(880–)	Adelaide de Toulouse	(216, 343, 956, 448, 1689)
18	Daughter	(925–989)	Senegonde de Rouergue	(107, 1556)
19	Daughter	(945–1011)	Adelaide de Rouergue	(216, 1556, 107, 711)
20	Son	(979–1038)	Bernard I Roger de Couserans	(216, 2, 107, 1689, 115)
21	Son	(1016–1077)	Bernard II of Bigorre	(566, 711, 107)
22	Daughter	(1035–1088)	Stephanie of Bigorre	(216, 193, 2, 115, 107)
23	Daughter	(1070–1135)	Gisele of Burgundy	(172, 216, 2, 115, 107)
24	Daughter	(1092–1154)	Adelaide de Savoy	(135, 172, 216, 224, 2)
25	Son	(1119–1180)	Louis VII of France	(78, 216, 239, 2, 35)
26	Daughter	(1150–1197)	Alice of France	(216, 109, 108, 260, 244)
27	Daughter	(1170–1230)	Marguerite of Blois	(150, 216, 221, 244, 258)
28	Daughter	(1194–1231)	Beatrix of Burgundy	(150, 244, 258)
29	Daughter	(1210–1270)	Beatrix of Meran	(150, 193, 224, 130)
30	Son	(1230–1283)	Hermann III of Orlamunde	(150, 224, 193, 130)
31	Daughter	(1260–1333)	Elisabet von Ballenstedt	(150, 193, 224, 130, 1191)
32	Daughter	(1286–1359)	Elisabet of Lobdeburg-Arnshaugk	(150, 311, 1393, 130)
33	Son	(1310–1349)	Friedrich II of Meissen	(150, 311, 1393, 130)
34	Daughter	(1329–1375)	Elisabeth of Meissen	(150, 311, 1393, 130)
35	Son	(1371–1440)	Friedrich I von Brandenburg	(78, 150, 194, 42, 71)
36	Son	(1414–1486)	Albrecht III Achilles von Brandenburg	(78, 193, 204, 216, 206)
37	Son	(1455–1499)	Johann IV Cicero von Brandenburg	(78, 194, 150, 42, 106)

(continued)

Table B.7 (continued)

Gen	Relation	(Birth–Death)	Person	(Source numbers)
38	Son	(1484–1535)	Joachim I Nestor of Brandenburg	(78, 150, 194, 193, 204)
39	Daughter	(1511–1577)	Margarete of Brandenburg	(193, 150, 129, 258, 589)
40	Son	(1536–1586)	Joachim Ernst II of Anhalt-Zerbst	(150, 193, 194, 78, 117)
41	Daughter	(1563–1607)	Elisabeth of Anhalt-Zerbst	(150, 193, 589, 119, 566)
42	Daughter	(1582–1616)	Magdalena von Brandenburg	(150, 193, 589, 566)
43	Daughter	(1601–1659)	Anna Eleonore of Hesse-Darmstadt	(51, 150, 193, 258, 589)
44	Son	(1629–1698)	Ernst Ausgustus of Brunswick-Luneburg	(78, 51, 150, 163, 52)
45	Son	(1660–1727)	George I Louis of Hanover	(78, 51, 150, 170, 163)
46	Son	(1683–1760)	George II August of Hanover	(78, 51, 144, 150, 193)
47	Daughter	(1723–1772)	Mary of Hanover	(78, 51, 144, 150, 193)
48	Son	(1747–1837)	Frederick III of Hesse-Cassel	(78, 144, 150, 193, 194)
49	Son	(1787–1867)	William of Hesse-Rumpenheim	(78, 193, 216, 194, 224)
50	Daughter	(1817–1898)	Louise Wilhelmina of Hesse-Cassel	(78, 216, 193, 194, 224)
51	Son	(1845–1913)	George I of the Hellenes	(78, 216, 194, 224, 46)
52	Son	(1882–1944)	Andrew of Greece and Denmark	(78, 150, 216, 193, 224)
53	Son	(1921–2021)	Philip Mountbatten	(78, 216, 224, 46, 52)
54	Son	(1948–)	Charles Philip Mountbatten-Windsor	(78, 216, 224, 46, 73)
55	Son	(1982–)	William Arthur Mountbatten-Windsor	(216, 95, 112, 113, 223)
56	Son	(2013–)	George Alexander Mountbatten-Windsor	(493, 1318)

Table B.8 Lineage of maximal ascent to Cunedda the Great from George Alexander Louis Mountbatten-Windsor

Gen	Relation	(Birth–Death)	Person	(Source numbers)
0	Start	(409–460)	Cunedda the Great	(52, 66, 73, 95, 143)
1	Son	(442–)	Ceredig of Gwynedd	(1, 52, 73, 95, 328)
2	Son	(460–)	Corwyn of Gwynedd	(1, 52, 95, 328)
3	Daughter	(481–)	D'Anaumide verch Teithfalit	(303, 328, 95)
4	Daughter	(530–)	Alienor of the Bretons	(1)
5	Son	(556–)	Withur de Leon d'Acqs	(964)
6	Son	(572–)	Ausoch de Leon d'Acqs	(1, 303, 493)
7	Daughter	(587–)	Pritelle de Leon d'Acqs	(1, 328, 493, 1689)
8	Son	(602–658)	Judicael II de Cornouaille	(1, 95, 107, 303, 328)
9	Son	(632–711)	Gradlon Flam de Cornouaille	(956, 1689, 711)
10	Daughter	(660–)	Cheronnoge Congare de Cornouaille	(1689)
11	Son	(680–720)	Riwallon II de Poher de Bors	(1154, 711)
12	Daughter	(720–)	Marmoec de Poher	(1154, 711)
13	Son	(765–812)	Erispoe I Ernoch de Poher	(1641, 1689, 1154)
14	Son	(790–851)	Nominoe de Vannes	(1, 52, 95, 107, 328)
15	Son	(815–857)	Erispoe II de Vannes	(1, 52, 95, 107, 1689)
16	Daughter	(830–880)	Lotitia de Poher	(328, 1154, 1689, 107)
17	Son	(860–888)	Berenger Judicael I of Rennes	(216, 328, 1689, 107)
18	Daughter	(883–925)	Papie de Bayeux	(216, 2, 95, 115, 107)
19	Daughter	(917–969)	Gerloc of Normandy	(216, 2, 95, 115, 107)
20	Daughter	(945–1006)	Adelais of Aquitaine	(216, 2, 30, 84, 95)
21	Son	(972–1031)	Robert II the Pious	(175, 216, 2, 30, 52)
22	Daughter	(1009–1079)	Adelaide of France	(216, 2, 35, 52, 114)
23	Son	(1035–1093)	Robert I of Flanders	(150, 216, 193, 35, 106)
24	Daughter	(1070–1117)	Gertrude of Flanders	(216, 193, 106, 114, 115)
25	Daughter	(1087–)	Gisela of Lorraine	(193, 1709, 107, 382)
26	Daughter	(1115–1147)	Agnes von Saarbrucken	(150, 382, 130)
27	Daughter	(1135–1191)	Jutta of Swabia	(150, 130)

(continued)

Table B.8 (continued)

Gen	Relation	(Birth–Death)	Person	(Source numbers)
28	Son	(1155–1217)	Hermann I of Thuringia	(150, 193, 224, 130)
29	Daughter	(1198–1244)	Irmgard of Thuringia	(193, 150, 130)
30	Daughter	(1215–1277)	Jutta of Anhalt	(193, 194, 130)
31	Son	(1235–1291)	Heinrich I of Werle-Gustrow	(51, 150, 216, 194, 130)
32	Daughter	(1268–1312)	Richiza of Werle-Gustrow	(51, 150, 194, 1393, 130)
33	Son	(1304–1369)	Magnus I of Brunswick	(51, 150, 205, 193, 224)
34	Son	(1328–1373)	Magnus II of Brunswick	(51, 150, 194, 193, 224)
35	Son	(1355–1416)	Henry II of Brunswick-Luneburg	(51, 150, 193, 130, 311)
36	Daughter	(1388–1442)	Catharina of Brunswick-Luneburg	(150, 130, 311, 1393)
37	Son	(1412–1464)	Friedrich II of Saxony	(150, 78, 106, 119, 311)
38	Son	(1441–1486)	Ernst I of Saxony	(51, 150, 194, 78, 106)
39	Daughter	(1461–1521)	Christine of Saxony	(150, 194, 113, 119, 589)
40	Daughter	(1485–1555)	Elizabeth of Denmark	(150, 194, 119, 258, 1393)
41	Daughter	(1511–1577)	Margarete of Brandenburg	(193, 150, 129, 258, 589)
42	Son	(1536–1586)	Joachim Ernst II of Anhalt-Zerbst	(150, 193, 194, 78, 117)
43	Daughter	(1563–1607)	Elisabeth of Anhalt-Zerbst	(150, 193, 589, 119, 566)
44	Daughter	(1582–1616)	Magdalena von Brandenburg	(150, 193, 589, 566)
45	Daughter	(1601–1659)	Anna Eleonore of Hesse-Darmstadt	(51, 150, 193, 258, 589)
46	Son	(1629–1698)	Ernst Ausgustus of Brunswick-Luneburg	(78, 51, 150, 163, 52)
47	Son	(1660–1727)	George I Louis of Hanover	(78, 51, 150, 170, 163)
48	Son	(1683–1760)	George II August of Hanover	(78, 51, 144, 150, 193)
49	Daughter	(1723–1772)	Mary of Hanover	(78, 51, 144, 150, 193)
50	Son	(1747–1837)	Frederick III of Hesse-Cassel	(78, 144, 150, 193, 194)

(continued)

Table B.8 (continued)

Gen	Relation	(Birth–Death)	Person	(Source numbers)
51	Son	(1787–1867)	William of Hesse-Rumpenheim	(78, 193, 216, 194, 224)
52	Daughter	(1817–1898)	Louise Wilhelmina of Hesse-Cassel	(78, 216, 193, 194, 224)
53	Son	(1845–1913)	George I of the Hellenes	(78, 216, 194, 224, 46)
54	Son	(1882–1944)	Andrew of Greece and Denmark	(78, 150, 216, 193, 224)
55	Son	(1921–2021)	Philip Mountbatten	(78, 216, 224, 46, 52)
56	Son	(1948–)	Charles Philip Mountbatten-Windsor	(78, 216, 224, 46, 73)
57	Son	(1982–)	William Arthur Mountbatten-Windsor	(216, 95, 112, 113, 223)
58	Son	(2013–)	George Alexander Mountbatten-Windsor	(493, 1318)

Table B.9 Lineage of maximal ascent to Clodgar von Therouanne from George Alexander Louis Mountbatten-Windsor

Gen	Relation	(Birth–Death)	Person	(Source numbers)
0	Start	(445–)	Clodgar von Therouanne	(1154, 1689, 1548)
1	Daughter	(510–)	Mathilde von Boulogne	(1154, 1548)
2	Son	(535–570)	Richard de Ponthieu	(1154, 1548)
3	Son	(560–614)	Ricmar d'Artois	(1683, 1564)
4	Son	(575–641)	Aega de Hesbaye	(51, 1281, 1689, 22, 328)
5	Daughter	(600–678)	Siagree of France	(107, 328, 22, 1689)
6	Daughter	(630–670)	Hymnegilde de Treves	(52, 15, 1, 1517, 1689)
7	Daughter	(647–)	Berswinde of Autun	(107, 227, 1689, 22)
8	Son	(672–722)	Adalbert de Alsace	(107, 328, 1689, 22)
9	Son	(704–767)	Luitfried I Hugues d'Alsace	(1689, 1564)
10	Son	(733–779)	Girard I of Paris	(107, 227, 328, 343, 711)
11	Daughter	(779–839)	Aba de Morvois	(216, 193, 115, 107, 303)
12	Daughter	(800–851)	Ermengarde of Alsace	(150, 193, 30, 31, 115)
13	Son	(835–869)	Lothair II of Lorraine	(51, 150, 216, 30, 107)
14	Daughter	(852–907)	Gisela of Lorraine	(150, 216, 107, 328, 1689)
15	Daughter	(868–917)	Ragnhildis von Friesland	(150, 216, 107)

(continued)

Table B.9 (continued)

Gen	Relation	(Birth–Death)	Person	(Source numbers)
16	Daughter	(896–968)	Matilda von Ringleheim	(51, 150, 2, 115, 107)
17	Daughter	(913–984)	Gerberge of Saxony	(51, 144, 150, 216, 193)
18	Daughter	(943–981)	Mathilde of France	(150, 216, 2, 30, 107)
19	Daughter	(965–1017)	Gerberga of Upper Burgundy	(150, 194, 115, 107, 181)
20	Daughter	(990–1043)	Gisela of Swabia	(150, 115, 107, 293, 311)
21	Son	(1017–1056)	Henry III the Mild	(216, 150, 2, 106, 114)
22	Son	(1050–1106)	Henry IV of Holy Roman Empire	(150, 216, 2, 35, 96)
23	Daughter	(1073–1143)	Agnes of Franconia	(150, 2, 35, 106, 115)
24	Daughter	(1088–1110)	Hedwig-Eilike Hohenstaufen	(150, 244, 1376)
25	Daughter	(1103–1170)	Heilika von Lengenfeld-Hopfenohe	(150, 130, 244, 328, 566)
26	Daughter	(1130–1174)	Hedwig von Wittelsbach	(150, 244)
27	Son	(1159–1204)	Berthold VI of Meran	(216, 150, 115, 107, 130)
28	Daughter	(1181–1213)	Gertrude von Meran	(150, 216, 2, 95, 113)
29	Son	(1206–1270)	Bela IV of Hungary	(150, 216, 2, 62, 95)
30	Son	(1239–1272)	Stephen V of Hungary	(216, 2, 106, 113, 115)
31	Daughter	(1257–1323)	Marie Arpad of Hungary	(216, 2, 100, 106, 113)
32	Daughter	(1273–1299)	Marguerite de Naples	(216, 2, 69, 84, 90)
33	Daughter	(1293–1342)	Jeanne of Valois	(216, 2, 95, 115, 107)
34	Daughter	(1310–1356)	Marguerite of Hainault	(216, 150, 106, 260, 1393)
35	Son	(1336–1404)	Albert I of Bavaria	(150, 216, 224, 239, 206)
36	Daughter	(1377–1410)	Joanna of Bavaria	(150, 1393, 63)
37	Son	(1397–1439)	Albert II of Austria	(150, 216, 224, 63, 106)
38	Daughter	(1436–1505)	Elizabeth of Austria	(150, 216, 113, 119, 258)
39	Daughter	(1464–1512)	Sofia of Poland	(216, 150, 258, 589)
40	Son	(1481–1527)	Kasimir of Brandenburg-Bayreuth	(193, 150, 589)
41	Daughter	(1519–1567)	Maria of Brandenburg-Bayreuth	(150, 589)
42	Daughter	(1544–1592)	Dorothea Susanna of Simmern	(150, 311, 589)
43	Son	(1570–1605)	Johann III of Saxe-Weimar	(150, 193, 78, 311, 589)

(continued)

Table B.9 (continued)

Gen	Relation	(Birth–Death)	Person	(Source numbers)
44	Son	(1601–1675)	Ernst I of Saxe-Gotha	(193, 150, 78, 258, 311)
45	Son	(1646–1691)	Friedrich I of Saxe-Gotha	(150, 194, 193, 78, 311)
46	Daughter	(1670–1728)	Anna Sophia of Saxe-Gotha	(150, 194, 852)
47	Daughter	(1700–1780)	Anna Sophia von Schwarzburg-Rudolstadt	(78, 194, 150, 589, 852)
48	Daughter	(1731–1810)	Charlotte Sophia of Saxe-Coburg-Saalfeld	(150, 194, 78, 589, 852)
49	Son	(1756–1837)	Friedrich Franz of Mecklenburg-Schwerin	(150, 194, 78, 311, 589)
50	Daughter	(1779–1801)	Luise Charlotte of Mecklenburg-Schwerin	(150, 194, 78, 311, 589)
51	Daughter	(1800–1831)	Dorothy Luise of Saxe-Gotha-Altenburg	(78, 150, 311, 589, 852)
52	Son	(1819–1861)	Francis Albert of Saxe-Coburg-Gotha	(78, 150, 216, 193, 194)
53	Daughter	(1843–1878)	Alice Maud Mary of Saxe-Coburg-Gotha	(78, 216, 193, 224, 46)
54	Daughter	(1863–1950)	Victoria of Hesse and the Rhine	(78, 193, 224, 239, 46)
55	Daughter	(1885–1969)	Victoria Alice Elizabeth of Battenberg	(78, 216, 224, 239, 46)
56	Son	(1921–2021)	Philip Mountbatten	(78, 216, 224, 46, 52)
57	Son	(1948–)	Charles Philip Mountbatten-Windsor	(78, 216, 224, 46, 73)
58	Son	(1982–)	William Arthur Mountbatten-Windsor	(216, 95, 112, 113, 223)
59	Son	(2013–)	George Alexander Mountbatten-Windsor	(493, 1318)

Table B.10 Lineage of maximal ascent to Charibert II of Neustria from George Alexander Louis Mountbatten-Windsor

Gen	Relation	(Birth–Death)	Person	(Source numbers)
0	Start	(565–636)	Charibert II Caribert of Neustria	(216, 1689, 107)
1	Son	(595–654)	Charibert Erlebert of Neustria	(1689, 107)
2	Son	(620–678)	Robert II Chrotbert Rupert II of Franks	(216, 1689)
3	Daughter	(642–)	Adelinde de Neustria	(1689, 1641)
4	Daughter	(660–)	Nanchtilde de Wormsgau	(1689)
5	Daughter	(680–746)	Gunilda d'Ascanie	(1689)
6	Daughter	(700–)	Herswinde of Saxony	(1689)
7	Daughter	(736–798)	Emma of Alemannien	(150, 107, 303, 328, 343)
8	Daughter	(757–783)	Hildegarde of Swabia	(51, 150, 30, 31, 52)
9	Son	(778–840)	Louis I the Pious	(51, 144, 150, 216, 2)
10	Son	(795–855)	Lothaire I of Italy	(144, 150, 175, 216, 193)
11	Son	(835–869)	Lothair II of Lorraine	(51, 150, 216, 30, 107)
12	Daughter	(852–907)	Gisela of Lorraine	(150, 216, 107, 328, 1689)
13	Daughter	(868–917)	Ragnhildis von Friesland	(150, 216, 107)
14	Daughter	(896–968)	Matilda von Ringleheim	(51, 150, 2, 115, 107)
15	Daughter	(913–984)	Gerberge of Saxony	(51, 144, 150, 216, 193)
16	Daughter	(943–981)	Mathilde of France	(150, 216, 2, 30, 107)
17	Daughter	(965–1017)	Gerberga of Upper Burgundy	(150, 194, 115, 107, 181)
18	Daughter	(990–1043)	Gisela of Swabia	(150, 115, 107, 293, 311)
19	Son	(1017–1056)	Henry III the Mild	(216, 150, 2, 106, 114)
20	Son	(1050–1106)	Henry IV of Holy Roman Empire	(150, 216, 2, 35, 96)
21	Daughter	(1073–1143)	Agnes of Franconia	(150, 2, 35, 106, 115)
22	Daughter	(1088–1110)	Hedwig-Eilike Hohenstaufen	(150, 244, 1376)
23	Daughter	(1103–1170)	Heilika von Lengenfeld-Hopfenohe	(150, 130, 244, 328, 566)
24	Daughter	(1130–1174)	Hedwig von Wittelsbach	(150, 244)
25	Son	(1159–1204)	Berthold VI of Meran	(216, 150, 115, 107, 130)
26	Daughter	(1181–1213)	Gertrude von Meran	(150, 216, 2, 95, 113)
27	Son	(1206–1270)	Bela IV of Hungary	(150, 216, 2, 62, 95)

(continued)

Table B.10 (continued)

Gen	Relation	(Birth–Death)	Person	(Source numbers)
28	Son	(1239–1272)	Stephen V of Hungary	(216, 2, 106, 113, 115)
29	Daughter	(1257–1323)	Marie Arpad of Hungary	(216, 2, 100, 106, 113)
30	Daughter	(1273–1299)	Marguerite de Naples	(216, 2, 69, 84, 90)
31	Daughter	(1293–1342)	Jeanne of Valois	(216, 2, 95, 115, 107)
32	Daughter	(1310–1356)	Marguerite of Hainault	(216, 150, 106, 260, 1393)
33	Son	(1336–1404)	Albert I of Bavaria	(150, 216, 224, 239, 206)
34	Daughter	(1377–1410)	Joanna of Bavaria	(150, 1393, 63)
35	Son	(1397–1439)	Albert II of Austria	(150, 216, 224, 63, 106)
36	Daughter	(1436–1505)	Elizabeth of Austria	(150, 216, 113, 119, 258)
37	Daughter	(1464–1512)	Sofia of Poland	(216, 150, 258, 589)
38	Son	(1481–1527)	Kasimir of Brandenburg-Bayreuth	(193, 150, 589)
39	Daughter	(1519–1567)	Maria of Brandenburg-Bayreuth	(150, 589)
40	Daughter	(1544–1592)	Dorothea Susanna of Simmern	(150, 311, 589)
41	Son	(1570–1605)	Johann III of Saxe-Weimar	(150, 193, 78, 311, 589)
42	Son	(1601–1675)	Ernst I of Saxe-Gotha	(193, 150, 78, 258, 311)
43	Son	(1646–1691)	Friedrich I of Saxe-Gotha	(150, 194, 193, 78, 311)
44	Daughter	(1670–1728)	Anna Sophia of Saxe-Gotha	(150, 194, 852)
45	Daughter	(1700–1780)	Anna Sophia von Schwarzburg-Rudolstadt	(78, 194, 150, 589, 852)
46	Daughter	(1731–1810)	Charlotte Sophia of Saxe-Coburg-Saalfeld	(150, 194, 78, 589, 852)
47	Son	(1756–1837)	Friedrich Franz of Mecklenburg-Schwerin	(150, 194, 78, 311, 589)
48	Daughter	(1779–1801)	Luise Charlotte of Mecklenburg-Schwerin	(150, 194, 78, 311, 589)
49	Daughter	(1800–1831)	Dorothy Luise of Saxe-Gotha-Altenburg	(78, 150, 311, 589, 852)
50	Son	(1819–1861)	Francis Albert of Saxe-Coburg-Gotha	(78, 150, 216, 193, 194)
51	Daughter	(1843–1878)	Alice Maud Mary of Saxe-Coburg-Gotha	(78, 216, 193, 224, 46)
52	Daughter	(1863–1950)	Victoria of Hesse and the Rhine	(78, 193, 224, 239, 46)

(continued)

Table B.10 (continued)

Gen	Relation	(Birth–Death)	Person	(Source numbers)
53	Daughter	(1885–1969)	Victoria Alice Elizabeth of Battenberg	(78, 216, 224, 239, 46)
54	Son	(1921–2021)	Philip Mountbatten	(78, 216, 224, 46, 52)
55	Son	(1948–)	Charles Philip Mountbatten-Windsor	(78, 216, 224, 46, 73)
56	Son	(1982–)	William Arthur Mountbatten-Windsor	(216, 95, 112, 113, 223)
57	Son	(2013–)	George Alexander Mountbatten-Windsor	(493, 1318)

Table B.11 Lineage of maximal ascent to Charlemagne from George Alexander Louis Mountbatten-Windsor

Gen	Relation	(Birth–Death)	Person	(Source numbers)
0	Start	(747–814)	Charlemagne the Great	(51, 144, 150, 1, 2)
1	Son	(778–840)	Louis I the Pious	(51, 144, 150, 216, 2)
2	Son	(795–855)	Lothaire I of Italy	(144, 150, 175, 216, 193)
3	Son	(835–869)	Lothair II of Lorraine	(51, 150, 216, 30, 107)
4	Daughter	(852–907)	Gisela of Lorraine	(150, 216, 107, 328, 1689)
5	Daughter	(868–917)	Ragnhildis von Friesland	(150, 216, 107)
6	Daughter	(896–968)	Matilda von Ringleheim	(51, 150, 2, 115, 107)
7	Daughter	(913–984)	Gerberge of Saxony	(51, 144, 150, 216, 193)
8	Daughter	(943–981)	Mathilde of France	(150, 216, 2, 30, 107)
9	Daughter	(965–1017)	Gerberga of Upper Burgundy	(150, 194, 115, 107, 181)
10	Daughter	(990–1043)	Gisela of Swabia	(150, 115, 107, 293, 311)
11	Son	(1017–1056)	Henry III the Mild	(216, 150, 2, 106, 114)
12	Son	(1050–1106)	Henry IV of Holy Roman Empire	(150, 216, 2, 35, 96)
13	Daughter	(1073–1143)	Agnes of Franconia	(150, 2, 35, 106, 115)
14	Daughter	(1088–1110)	Hedwig-Eilike Hohenstaufen	(150, 244, 1376)
15	Daughter	(1103–1170)	Heilika von Lengenfeld-Hopfenohe	(150, 130, 244, 328, 566)
16	Daughter	(1130–1174)	Hedwig von Wittelsbach	(150, 244)
17	Son	(1159–1204)	Berthold VI of Meran	(216, 150, 115, 107, 130)

(continued)

Table B.11 (continued)

Gen	Relation	(Birth–Death)	Person	(Source numbers)
18	Daughter	(1181–1213)	Gertrude von Meran	(150, 216, 2, 95, 113)
19	Son	(1206–1270)	Bela IV of Hungary	(150, 216, 2, 62, 95)
20	Son	(1239–1272)	Stephen V of Hungary	(216, 2, 106, 113, 115)
21	Daughter	(1257–1323)	Marie Arpad of Hungary	(216, 2, 100, 106, 113)
22	Daughter	(1273–1299)	Marguerite de Naples	(216, 2, 69, 84, 90)
23	Daughter	(1293–1342)	Jeanne of Valois	(216, 2, 95, 115, 107)
24	Daughter	(1310–1356)	Marguerite of Hainault	(216, 150, 106, 260, 1393)
25	Son	(1336–1404)	Albert I of Bavaria	(150, 216, 224, 239, 206)
26	Daughter	(1377–1410)	Joanna of Bavaria	(150, 1393, 63)
27	Son	(1397–1439)	Albert II of Austria	(150, 216, 224, 63, 106)
28	Daughter	(1436–1505)	Elizabeth of Austria	(150, 216, 113, 119, 258)
29	Daughter	(1464–1512)	Sofia of Poland	(216, 150, 258, 589)
30	Son	(1481–1527)	Kasimir of Brandenburg-Bayreuth	(193, 150, 589)
31	Daughter	(1519–1567)	Maria of Brandenburg-Bayreuth	(150, 589)
32	Daughter	(1544–1592)	Dorothea Susanna of Simmern	(150, 311, 589)
33	Son	(1570–1605)	Johann III of Saxe-Weimar	(150, 193, 78, 311, 589)
34	Son	(1601–1675)	Ernst I of Saxe-Gotha	(193, 150, 78, 258, 311)
35	Son	(1646–1691)	Friedrich I of Saxe-Gotha	(150, 194, 193, 78, 311)
36	Daughter	(1670–1728)	Anna Sophia of Saxe-Gotha	(150, 194, 852)
37	Daughter	(1700–1780)	Anna Sophia von Schwarzburg-Rudolstadt	(78, 194, 150, 589, 852)
38	Daughter	(1731–1810)	Charlotte Sophia of Saxe-Coburg-Saalfeld	(150, 194, 78, 589, 852)
39	Son	(1756–1837)	Friedrich Franz of Mecklenburg-Schwerin	(150, 194, 78, 311, 589)
40	Daughter	(1779–1801)	Luise Charlotte of Mecklenburg-Schwerin	(150, 194, 78, 311, 589)
41	Daughter	(1800–1831)	Dorothy Luise of Saxe-Gotha-Altenburg	(78, 150, 311, 589, 852)
42	Son	(1819–1861)	Francis Albert of Saxe-Coburg-Gotha	(78, 150, 216, 193, 194)

(continued)

Table B.11 (continued)

Gen	Relation	(Birth–Death)	Person	(Source numbers)
43	Daughter	(1843–1878)	Alice Maud Mary of Saxe-Coburg-Gotha	(78, 216, 193, 224, 46)
44	Daughter	(1863–1950)	Victoria of Hesse and the Rhine	(78, 193, 224, 239, 46)
45	Daughter	(1885–1969)	Victoria Alice Elizabeth of Battenberg	(78, 216, 224, 239, 46)
46	Son	(1921–2021)	Philip Mountbatten	(78, 216, 224, 46, 52)
47	Son	(1948–)	Charles Philip Mountbatten-Windsor	(78, 216, 224, 46, 73)
48	Son	(1982–)	William Arthur Mountbatten-Windsor	(216, 95, 112, 113, 223)
49	Son	(2013–)	George Alexander Mountbatten-Windsor	(493, 1318)

Appendix C: Sample Lineages for Linking Historically Influential Persons to the Unifying Ancestry

In aggregate the lineages to influential persons in the Research Genealogy comprise a Genealogical History of the Modern World, showing the familial relationships between persons involved in historical events.

Sample lineages of Maximal Ascent are provided that span the period from 1035 to 1874 for persons referenced by both Hart "The 100" and Palmer "History of the Modern World". Lineages of maximal ascent to Charlemagne are shown for each notable person. Each generation of the lineage is annotated with up to five source identifiers. Appendix D provides an ordered list of all the referenced source identifiers. In each table:

- Gen is the generation number
- Relation is the familial relationship from the previous person in the list.
- Up to 5 sources are listed for each person

Lineages for the notable historical person are shown in the following tables. For each person, the birth and death years are listed. Note that all the notable persons are descendants of Charlemagne and almost all the Progenitors listed in Table 19. The exception is Odo of Chatillon who is only a descendant of Charibert II of Neustria, Clodgar von Therouanne, Potitus Photaighe of Ireland, and Danp of Denmark.

Notable person	Table	(Birth–Death)
Odo of Chatillon	Table C.1	(1035–1099)
Thomas d'Aquino	Table C.2	(1225–1273)
Louis XI of France	Table C.3	(1423–1483)
Ivan III the Great	Table C.4	(1440–1505)
Fernao de Magalhaes	Table C.5	(1480–1521)
Francis Bacon	Table C.6	(1560–1626)
William Shakespeare	Table C.7	(1564–1616)
John Locke	Table C.8	(1632–1704)
Antoni van Leeuwenhoek	Table C.9	(1632–1723)

© The Editor(s) (if applicable) and The Author(s), under exclusive license to Springer Nature Switzerland AG 2023
R. W. Moore, *Trustworthy Communications and Complete Genealogies*, Synthesis Lectures on Information Concepts, Retrieval, and Services,
https://doi.org/10.1007/978-3-031-16836-9

Notable person	Table	(Birth–Death)
Isaac Newton	Table C.10	(1642–1727)
Benjamin Franklin	Table C.11	(1706–1790)
Adam Smith	Table C.12	(1723–1790)
Robert Clive	Table C.13	(1726–1774)
James Watt	Table C.14	(1736–1819)
Eli Whitney	Table C.15	(1765–1815)
Robert Fulton	Table C.16	(1726–1774)
Thomas Robert Malthus	Table C.17	(1766–1834)
Napoleon I Buonaparte	Table C.18	(1769–1821)
Simon Jose de Bolivar	Table C.19	(1783–1830)
Victor Hugo	Table C.20	(1802–1885)
Charles Robert Darwin	Table C.21	(1809–1882)
Victor Emmanuel II of Sardinia	Table C.22	(1820–1878)
James Clerk Maxwell	Table C.23	(1831–1879)
Thomas Alva Edison	Table C.24	(1847–1931)
Max Karl Planck	Table C.25	(1858–1947)
Henry Ford	Table C.26	(1863–1947)
Wilbur Wright	Table C.27	(1867–1912)

The lineages take between 9 and 35 generations to connect to the lineage of maximal ascent from Prince George of Cambridge to Charlemagne. Most of the lineages take at least 23 generations to connect. Each notable person needs between 1 and 17 generations to connect to the closest common ancestor with Prince George. Connecting to the lineage of maximal ascent takes more generations than connecting to the common ancestor.

– Odo of Chatillon is the 7th Great-Grandson of Charlemagne and has 4 descents from Charlemagne. Note that the shortest lineage from Odo to Charlemagne has one fewer generation than the lineage of maximal ascent. Odo is the 25th Great-GrandUncle of Prince George of Cambridge. Odo was Pope Urban II, known for initiating the Crusades. The lineage of maximal ascent goes through France. It takes 9 generations to connect to the lineage of maximal ascent from Prince George of Cambridge to Charlemagne. The connection occurs with Louis I the Pious, born in 778. The number of generations from Odo to the closest common ancestor with Prince George is 1.

– Thomas d'Aquino is the 12th Great-Grandson of Charlemagne and has 145 descents from Charlemagne. Thomas is the 2nd cousin 22 removes of Prince George of Cambridge. Thomas inspired modern philosophical areas related to ethics, metaphysics, and political theory. The lineage goes from France through Italy to Germany before ending in Sicily. It takes 11 generations to connect to the lineage of maximal ascent from Prince George of Cambridge to Charlemagne. The connection occurs with Matilde von Ringleheim, born in 896. The number of generations from Thomas to the closest common ancestor of Prince George is 3.

- King Louis XI of France is the 17th Great-Grandson of Charlemagne and has 110,122 ascents to Charlemagne. Louis is the 15th Great-Uncle of Prince George of Cambridge. Louis XI signed the Treaty of Picquigny with England, ending the Hundred Years' War. The lineage goes through Italy, France, and Spain before ending in France. It takes 20 generations to connect to the lineage of maximal ascent from Prince George of Cambridge to Charlemagne. The connection occurs with Gerberge of Saxony, born in 913. The number of generations from King Louis XI to the closest common ancestor of Prince George is 1.

- Tsar Ivan III Vasilievitch the Great is the 19th Great-Grandson of Charlemagne and has 3135 ascents to Charlemagne. Ivan is the 3rd cousin 14 removes of Prince George of Cambridge. Ivan's victory in 1480 over the Great Horde ended the domination of the Tatars over Russia. The lineage goes through France and Germany before ending in Russia. It takes 15 generations to connect to the lineage of maximal ascent from Prince George of Cambridge to Charlemagne. The connection occurs with Agnes of Franconia, born in 1073. The number of generations from Ivan III to the closest common ancestor of Prince George is 4.

- Fernao de Magalhaes (Ferdinand Magellan) is the 19th Great-Grandson of Charlemagne and has 760 ascents to Charlemagne. Ferdinand is the half 7th cousin 16 removes of Prince George of Cambridge. Ferdinand led the 1519 Spanish expedition from the Atlantic across the Pacific Ocean to the East Indies. The lineage goes through Italy, France, and Spain before ending in Portugal. It takes 22 generations to connect to the lineage of maximal ascent from Prince George of Cambridge to Charlemagne. The connection occurs with Lothair II of Lorraine, born in 835. The number of generations from Ferdinand to the closest common ancestor of Prince George is 8.

- Francis Bacon is the 21st Great-Grandson of Charlemagne and has 19,130 ascents to Charlemagne. Francis is the 1st cousin 14 removes of Prince George of Cambridge. Francis Bacon is the father of empiricism and the development of the scientific method. The lineage goes through France and Italy before ending in England. It takes 21 generations to connect to the lineage of maximal ascent from Prince George of Cambridge to Charlemagne. The connection occurs with Gerberge of Saxony, born in 913. The number of generations from Francis to the closest common ancestor of Prince George is 2.

- William Shakespeare is the 23rd Great-Grandson of Charlemagne and has 22,544 ascents to Charlemagne. Shakespeare is the 4th cousin 14 removes of Prince George of Cambridge. Shakespeare is known as the world's greatest dramatist. The lineage goes through France and Italy before ending in England. It takes 23 generations to connect to the lineage of maximal ascent from Prince George of Cambridge to Charlemagne. The connection occurs with Gerberge of Saxony, born in 913. The number of generations from William to the closest common ancestor of Prince George is 5.

- John Locke is the 24th Great-Grandson of Charlemagne and has 89,909 ascents to Charlemagne. Locke is the 7th cousin 8 removes of Prince George of Cambridge. Locke is known as the "Father of Liberalism", influencing the development of epistemology and political philosophy. The lineage goes through France, Italy, England, Spain, and Portugal before ending in England. It takes 25 generations to connect to the lineage of maximal ascent from Prince George of Cambridge to Charlemagne. The connection occurs with Gerberge of Saxony, born in 913. The number of generations from John to the closest common ancestor of Prince George is 8.

- Antoni van Leeuwenhoek is the 23rd Great-Grandson of Charlemagne and has 1986 ascents to Charlemagne. Antoni is the half 10th cousin 11 removes of Prince George of Cambridge. Leeuwenhoek is known as the "Father of Microbiology". He developed microscopes to observe microbes, muscle fibers, bacteria, red blood cells, and blood flow. The lineage goes through France, Italy, and France before ending in the Dutch Republic. It takes 21 generations to connect to the lineage of maximal ascent from Prince George of Cambridge to Charlemagne. The connection occurs with Gerberge of Saxony, born in 913. The number of generations from Antoni to the closest common ancestor of Prince George is 11.

- Isaac Newton is the 24th Great-Grandson of Charlemagne and has 310,246 ascents to Charlemagne. Newton is the 7th cousin 10 removes of Prince George of Cambridge. Newton formulated the laws of motion and universal gravitation, establishing classical mechanics. The lineage goes through France, Italy and France before ending in England. It takes 28 generations to connect to the lineage of maximal ascent from Prince George of Cambridge to Charlemagne. The connection occurs with Gerberge of Saxony, born in 913. The number of generations from Isaac to the closest common ancestor of Prince George is 8.

- Benjamin Franklin is the 26th Great-Grandson of Charlemagne and has 91,639 ascents to Charlemagne. Franklin is the 9th cousin 8 removes of Prince George of Cambridge. Franklin was an American Polymath who was active as a writer, scientist, inventor, statesman, diplomat, printer, publisher and political philosopher. Franklin invented the lightning rod, bifocals, and the Franklin stove. The lineage goes through France, Italy, and England before ending in the United States. It takes 29 generations to connect to the lineage of maximal ascent from Prince George of Cambridge to Charlemagne. The connection occurs with Gerberge of Saxony, born in 913. The number of generations from Benjamin to the closest common ancestor of Prince George is 10.

– Adam Smith is the 26th Great-Grandson of Charlemagne and has 69,312 ascents to Charlemagne. Smith is the 5th cousin 10 removes of Prince George of Cambridge. Smith is known as the "Father of Economics" and the "Father of Capitalism". The lineage goes through France and Italy before ending in Scotland. It takes 28 generations to connect to the lineage of maximal ascent from Prince George of Cambridge to Charlemagne. The connection occurs with Gerberge of Saxony, born in 913. The number of generations from Adam to the closest common ancestor of Prince George is 6.

– Robert Clive is the 28th Great-Grandson of Charlemagne and has 69,986 ascents to Charlemagne. Clive is the 8th cousin 7 removes of Prince George of Cambridge. Clive laid the foundation for the British rule of India through the use of puppet governments. The lineage goes through France and Italy before ending in England. It takes 28 generations to connect to the lineage of maximal ascent from Prince George of Cambridge to Charlemagne. The connection occurs with Gerberge of Saxony, born in 913. The number of generations from Robert to the closest common ancestor of Prince George is 9.

– James Watt is the 28th Great-Grandson of Charlemagne and has 19,940 ascents to Charlemagne. Watt is the half 8th cousin 8 removes of Prince George of Cambridge. Watt improved the performance of steam engines and developed the concept of horsepower. The lineage goes through France and Italy before ending in Scotland. It takes 36 generations to connect to the lineage of maximal ascent from Prince George of Cambridge to Charlemagne. The connection occurs with Louis I the Pious, born in 778. The number of generations from James to the closest common ancestor of Prince George is 9.

– Eli Whitney is the 27th Great-Grandson of Charlemagne and has 191,039 ascents to Charlemagne. Whitney is the 5th cousin 9 removes of Prince George of Cambridge. Whitney invented the cotton gin, hoping to reduce the use of slavery. The lineage goes through France, Italy, and England before ending in the United States. It takes 32 generations to connect to the lineage of maximal ascent from Prince George of Cambridge to Charlemagne. The connection occurs with Gerberge of Saxony, born in 913. The number of generations from Eli to the closest common ancestor of Prince George is 6.

– Robert Fulton is the 28th Great-Grandson of Charlemagne and has 135,457 ascents to Charlemagne. Fulton is the 7th cousin 9 removes of Prince George of Cambridge. Fulton developed the first commercially successful steamboat. The lineage goes through France, Germany, and England before ending in the United States. It takes 27 generations to connect to the lineage of maximal ascent from Prince George of Cambridge to Charlemagne. The connection occurs with Gerberge of Saxony, born in 913. The number of generations from Robert to the closest common ancestor of Prince George is 8.

– Thomas Robert Malthus is the 27th Great-Grandson of Charlemagne and has 27,692 ascents to Charlemagne. Malthus is the 8th cousin 10 removes of Prince George of Cambridge. Malthus was an economist and promoted the "Malthusian trap" in which improvements in food supply led to increased population instead of a higher standard of living. The lineage goes through France and Italy before ending in England. It takes 29 generations to connect to the lineage of maximal ascent from Prince George of Cambridge to Charlemagne. The connection occurs with Gergerbe of Saxony, born in 913. The number of generations from Thomas to the closest common ancestor of Prince George is 9.

– Emperor Napoleon I Buonaparte is the 28th Great-Grandson of Charlemagne and has 97 ascents to Charlemagne. Napoleon is the 12th cousin 9 removes of Prince George of Cambridge. Napoleon was Emperor of the French from 1804 to 1815 and is known as one of the greatest military commanders in history. The lineage goes through France and Italy, before ending in Corsica. It takes 26 generations to connect to the lineage of maximal ascent from Prince George of Cambridge to Charlemagne. The connection occurs with Gerberge of Saxony, born in 913. The number of generations from Napoleon to the closest common ancestor of Prince George is 13.

– Simon Bolivar is the 29th Great-Grandson of Charlemagne and has 68,502 ascents to Charlemagne. Bolivar is the 13th cousin 8 removes of Prince George of Cambridge. Bolivar is known as "El Libertador", freeing Colombia from Spanish rule. The lineage goes through France, Italy and Spain before ending in Venezuela. It takes 31 generations to connect to the lineage of maximal ascent from Prince George of Cambridge to Charlemagne. The connection occurs with Gerberge of Saxony, born in 913. The number of generations from Simon to the closest common ancestor of Prince George is 14.

– Victor Hugo is the 29th Great-Grandson of Charlemagne and has 8014 ascents to Charlemagne. Hugo is the 13th cousin 6 removes of Prince George of Cambridge. Hugo is known as one of the greatest French writers. The lineage goes through France and Italy before ending in France. It takes 30 generations to connect to the lineage of maximal ascent from Prince George of Cambridge to Charlemagne. The connection occurs with Gerberge of Saxony, born in 913. The number of generations from Victor to the closest common ancestor of Prince George is 14.

– Charles Darwin is the 28th Great-Grandson of Charlemagne and has 1,171,891 ascents to Charlemagne. Darwin is the 5th cousin 7 removes of Prince George of Cambridge. Darwin published the theory of evolution in 1859. The lineage goes through France and Italy before ending in England. It takes 30 generations to connect to the lineage of maximal ascent from Prince George of Cambridge to Charlemagne. The connection occurs with Gerberge of Saxony, born in 913. The number of generations from Charles to the closest common ancestor of Prince George is 6.

– Victor Emmanuel II of Sardinia is the 27th Great-Grandson of Charlemagne and has 364 million ascents to Charlemagne. Victor is the 5th cousin 5 removes of Prince George of Cambridge. Victor was the first king of a united Italy since the sixth century. The lineage goes through France, Italy, Hungary, and Austria before ending in Italy. It takes 18 generations to connect to the lineage of maximal ascent from Prince George of Cambridge to Charlemagne. The connection occurs with Albert I of Bavaria, born in 1336. The number of generations from Victor to the closest common ancestor of Prince George is 6.

– James Clerk Maxwell is the 31st Great-Grandson of Charlemagne and has 7354 ascents to Charlemagne. Maxwell is the 10th cousin 7 removes of Prince George of Cambridge. Maxwell developed the classical theory of electromagnetism which unified electricity, magnetism, and light. The lineage goes through France and Italy before ending in England. It takes 36 generations to connect to the lineage of maximal ascent from Prince George of Cambridge to Charlemagne. The connection occurs with Louis I the Pious, born in 778. The number of generations from James to the closest common ancestor of Prince George is 11.

– Thomas Edison is the 30th Great-Grandson of Charlemagne and has 82,765 ascents to Charlemagne. Edison is the 10th cousin 4 removes of Prince George of Cambridge. Edison invented the electric light bulb, motion picture camera, phonograph, and electric power generation. The lineage goes through France and Italy before ending in England. It takes 32 generations to connect to the lineage of maximal ascent from Prince George of Cambridge to Charlemagne. The connection occurs with Gerberge of Saxony, born in 913. The number of generations from Thomas to the closest common ancestor of Prince George is 11.

– Max Planck is the 30th Great-Grandson of Charlemagne and has 20,690 ascents to Charlemagne. Planck is the half 13th cousin 4 removes of Prince George of Cambridge. Planck developed Quantum Theory. The lineage goes through France and Italy before ending in Germany. It takes 23 generations to connect to the lineage of maximal ascent from Prince George of Cambridge to Charlemagne. The connection occurs with Gertrude von Meran, born in 1181. The number of generations from Max to the closest common ancestor of Prince George is 14.

– Henry Ford is the 30th Great-Grandson of Charlemagne and has 191,570 ascents to Charlemagne. Ford is the 11th cousin 5 removes of Prince George of Cambridge. Ford developed the assembly line for mass production. The lineage goes through France, Italy and England before ending in the United States. It takes 32 generations to connect to the lineage of maximal ascent from Prince George of Cambridge to Charlemagne. The connection occurs with Gerberge of Saxony, born in 913. The number of generations from Henry to the closest common ancestor of Prince George is 12.

– Wilbur Wright is the 30th Great-Grandson of Charlemagne and has 335,296 ascents to Charlemagne. Wright is the 7th cousin 6 removes of Prince George of Cambridge. With his brother Orville, Wilbur built the first successful motor-operated airplane. The lineage goes through France, Italy and England before ending in the United States. It takes 35 generations to connect to the lineage of maximal ascent from Prince George of Cambridge to Charlemagne. The connection occurs with Gerberge of Saxony, born in 913. The number of generations from Wilbur to the closest common ancestor of Prince George is 8.

Table C.1 Lineage of maximal ascent from Odo of Chatillon to Charlemagne

Gen	Relation	(Birth–Death)	Person	(Source numbers)
0	Start	(747–814)	Charlemagne the Great	(51, 144, 150, 1, 2)
1	Son	(778–840)	Louis I the Pious	(51, 144, 150, 216, 2)
2	Son	(823–877)	Charles II the Bald	(144, 195, 216, 150, 2)
3	Daughter	(844–870)	Judith of France	(150, 216, 2, 30, 95)
4	Son	(865–896)	Ralph of Cambrai	(216, 114, 107, 711, 1683)
5	Daughter	(893–967)	Bertha of Cambrai	(216, 342, 711, 1683, 107)
6	Son	(930–967)	Arnaud of Cambrai	(342, 107)
7	Daughter	(956–1001)	Gisele of Cambrai	(342, 107)
8	Son	(976–1044)	Milon of Chatillon	(342, 1680, 107)
9	Son	(1007–1076)	Gui I of Chatillon	(1680, 227, 107)
10	Son	(1035–1099)	Odo of Chatillon	(304, 1318, 1502, 328, 342)

Table C.2 Lineage of maximal ascent from Thomas d'Aquino to Charlemagne

Gen	Relation	(Birth–Death)	Person	(Source numbers)
0	Start	(747–814)	Charlemagne the Great	(51, 144, 150, 1, 2)
1	Son	(778–840)	Louis I the Pious	(51, 144, 150, 216, 2)
2	Son	(795–855)	Lothaire I of Italy	(144, 150, 175, 216, 193)
3	Son	(835–869)	Lothair II of Lorraine	(51, 150, 216, 30, 107)
4	Daughter	(852–907)	Gisela of Lorraine	(150, 216, 107, 328, 1689)
5	Daughter	(868–917)	Ragnhildis von Friesland	(150, 216, 107)
6	Daughter	(896–968)	Matilda von Ringleheim	(51, 150, 2, 115, 107)
7	Son	(912–973)	Otto I the Great	(51, 150, 216, 193, 2)
8	Son	(930–957)	Liudolf of Saxony	(150, 95, 132, 181, 30)
9	Daughter	(946–1014)	Richilde of Saxony	(51, 115, 107, 711, 1154)
10	Daughter	(980–1020)	Kunigunde of Ohningen	(150, 107)
11	Son	(998–1072)	Frederick von Buren	(150, 193, 107)
12	Son	(1020–1068)	Frederick von Buren	(150, 115, 107, 311, 244)
13	Son	(1050–1105)	Frederick I Hohenstaufen	(150, 35, 106, 115, 107)
14	Son	(1090–1147)	Frederick II Hohenstaufen	(51, 150, 193, 1, 2)
15	Daughter	(1137–)	Francoise von Hohenstaufen	(382)
16	Son	(1160–1245)	Landolfo V d'Aquino	(382)
17	Son	(1225–1273)	Thomas d'Aquino	(304, 1318, 1502, 382)

Table C.3 Lineage of maximal ascent from Louis XI of France to Charlemagne

Gen	Relation	(Birth–Death)	Person	(Source numbers)
0	Start	(747–814)	Charlemagne the Great	(51, 144, 150, 1, 2)
1	Son	(778–840)	Louis I the Pious	(51, 144, 150, 216, 2)
2	Son	(795–855)	Lothaire I of Italy	(144, 150, 175, 216, 193)
3	Son	(835–869)	Lothair II of Lorraine	(51, 150, 216, 30, 107)
4	Daughter	(852–907)	Gisela of Lorraine	(150, 216, 107, 328, 1689)
5	Daughter	(868–917)	Ragnhildis von Friesland	(150, 216, 107)
6	Daughter	(896–968)	Matilda von Ringleheim	(51, 150, 2, 115, 107)
7	Daughter	(913–984)	Gerberge of Saxony	(51, 144, 150, 216, 193)
8	Daughter	(930–973)	Aubree of Lorraine	(216, 193, 2, 115, 107)
9	Daughter	(950–1005)	Ermentrude de Roucy	(216, 2, 115, 107, 315)
10	Daughter	(995–1068)	Agnes of Burgundy	(150, 216, 2, 115, 107)
11	Son	(1022–1058)	William VII of Aquitaine	(216, 224, 193, 206, 2)
12	Daughter	(1058–1129)	Clementia of Poitiers	(216, 224, 193, 206, 115)
13	Daughter	(1088–1137)	Yolande of Wassemberg	(216, 206, 2, 115, 107)
14	Son	(1108–1171)	Baldwin IV of Hainault	(135, 216, 224, 2, 106)
15	Son	(1150–1195)	Baldwin V of Hainault	(3, 216, 2, 106, 115)
16	Daughter	(1170–1190)	Isabelle of Hainault	(216, 2, 52, 84, 113)
17	Son	(1187–1226)	Louis VIII of France	(135, 216, 2, 52, 66)
18	Son	(1215–1270)	Louis IX of France	(216, 193, 2, 52, 66)
19	Son	(1245–1285)	Philip III the Bold	(51, 216, 193, 2, 52)
20	Son	(1270–1325)	Charles III de Valois	(138, 161, 216, 224, 221)
21	Son	(1293–1350)	Philip VI of France	(216, 239, 52, 69, 76)
22	Son	(1319–1364)	Jean II of France	(216, 224, 150, 193, 52)
23	Daughter	(1344–1404)	Mary of France	(216, 193, 90, 324, 1393)
24	Daughter	(1365–1431)	Jolande of Bar	(193, 216, 1393)
25	Daughter	(1384–1443)	Yolande of Aragon	(216, 15, 324, 1393)
26	Daughter	(1404–1463)	Marie of Anjou	(216, 52, 69, 15, 113)
27	Son	(1423–1483)	Louis XI of France	(216, 239, 52, 69, 73)

Table C.4 Lineage of maximal ascent from Ivan III the Great to Charlemagne

Gen	Relation	(Birth–Death)	Person	(Source numbers)
0	Start	(747–814)	Charlemagne the Great	(51, 144, 150, 1, 2)
1	Son	(778–840)	Louis I the Pious	(51, 144, 150, 216, 2)
2	Son	(795–855)	Lothaire I of Italy	(144, 150, 175, 216, 193)
3	Son	(835–869)	Lothair II of Lorraine	(51, 150, 216, 30, 107)
4	Daughter	(852–907)	Gisela of Lorraine	(150, 216, 107, 328, 1689)
5	Daughter	(868–917)	Ragnhildis von Friesland	(150, 216, 107)
6	Daughter	(896–968)	Matilda von Ringleheim	(51, 150, 2, 115, 107)
7	Daughter	(913–984)	Gerberge of Saxony	(51, 144, 150, 216, 193)
8	Daughter	(943–981)	Mathilde of France	(150, 216, 2, 30, 107)
9	Daughter	(965–1017)	Gerberga of Upper Burgundy	(150, 194, 115, 107, 181)
10	Daughter	(990–1043)	Gisela of Swabia	(150, 115, 107, 293, 311)
11	Son	(1017–1056)	Henry III the Mild	(216, 150, 2, 106, 114)
12	Son	(1050–1106)	Henry IV of Holy Roman Empire	(150, 216, 2, 35, 96)
13	Daughter	(1073–1143)	Agnes of Franconia	(150, 2, 35, 106, 115)
14	Son	(1093–1152)	Conrad III Hohenstaufen	(150, 216, 1, 35, 76)
15	Daughter	(1113–)	(daughter) of Hohenstaufen	(150, 216, 130)
16	Son	(1131–1170)	Mstislav II Isyaslavitch of Novgorod	(216, 114, 130, 106)
17	Son	(1160–1205)	Roman Mstislavitch of Kiev	(216, 130, 258, 114)
18	Daughter	(1195–1241)	Mariya Romanovna of Galicia	(216, 258, 343, 130)
19	Daughter	(1211–1271)	Mariya Michailovna of Tschernigov	(216, 258, 130)
20	Son	(1231–1277)	Boris Vasilkovitch of Rostov	(216, 258, 130)

(continued)

Table C.4 (continued)

Gen	Relation	(Birth–Death)	Person	(Source numbers)
21	Son	(1253–1294)	Dimitrii Borisovitch of Rostov	(216, 1393, 130)
22	Daughter	(1280–1368)	Anna Dimitrievna of Rostov	(216, 1393, 130)
23	Son	(1301–1339)	Alexander I Michailovitch of Tver	(216, 59, 448, 1393, 130)
24	Daughter	(1329–1371)	Alexandrovna of Tver	(448)
25	Daughter	(1357–1418)	Anna Svyateslavna of Smolensk	(216)
26	Daughter	(1377–1453)	Sophia Witowtovna of Lithuania	(216, 113, 258, 48)
27	Son	(1415–1462)	Vasily II Vasilyevich of Moscow	(216, 59, 106, 113, 258)
28	Son	(1440–1505)	Ivan III Vasilievitch the Great	(216, 3, 48, 59, 76)

Table C.5 Lineage of maximal ascent from Fernao de Magalhaes to Charlemagne

Gen	Relation	(Birth–Death)	Person	(Source numbers)
0	Start	(747–814)	Charlemagne the great	(51, 144, 150, 1, 2)
1	Son	(778–840)	Louis I the Pious	(51, 144, 150, 216, 2)
2	Son	(795–855)	Lothaire I of Italy	(144, 150, 175, 216, 193)
3	Son	(835–869)	Lothair II of Lorraine	(51, 150, 216, 30, 107)
4	Daughter	(863–925)	Bertha of Lorraine	(51, 150, 216, 2, 107)
5	Daughter	(883–948)	Teutberga de Arles	(216, 107, 342, 724, 1680)
6	Daughter	(902–960)	Teutberge of Arles	(216, 30, 107, 181, 328)
7	Daughter	(920–965)	Constance of Provence, Arles, and	(216, 293, 315, 328, 342)
8	Son	(955–994)	William I of Provence	(216, 293, 311, 315, 328)
9	Daughter	(986–1032)	Constance of Toulouse	(175, 216, 2, 6, 52)
10	Son	(1011–1076)	Robert I of Burgundy	(216, 2, 35, 84, 106)
11	Daughter	(1050–1104)	Hildegarde of Burgundy	(216, 2, 115, 107, 303)
12	Son	(1071–1127)	William IX of Aquitaine	(195, 216, 2, 35, 106)

(continued)

Table C.5 (continued)

Gen	Relation	(Birth–Death)	Person	(Source numbers)
13	Daughter	(1095–1147)	Agnes of Aquitaine	(195, 216, 2, 113, 115)
14	Daughter	(1135–1172)	Petronilla I of Aragon	(216, 2, 104, 106, 113)
15	Daughter	(1160–1198)	Aldonza of Barcelona	(216, 2, 108, 113, 130)
16	Son	(1185–1223)	Alfonso II of Portugal	(216, 104, 108, 106, 113)
17	Son	(1210–1279)	Alfonso III of Portugal	(216, 104, 108, 106, 113)
18	Son	(1260–1310)	Afonso Diniz de Portugal	(566, 711, 260)
19	Son	(1305–1370)	Rodrigo Afonso de Sousa	(711)
20	Son	(1330–1386)	Goncalo Rodrigues de Sousa	(711)
21	Son	(1350–)	Jorge de Sousa Idanha	(711)
22	Daughter	(1375–)	Isabel de Sousa	(711)
23	Daughter	(1400–)	Violante Sousa	(711)
24	Son	(1433–1500)	Rodrigo de Magalhaes	(711, 1149)
25	Son	(1480–1521)	Fernao de Magalhaes	(711, 1318, 1502, 1149, 304)

Table C.6 Lineage of maximal ascent from Francis Bacon to Charlemagne

Gen	Relation	(Birth–Death)	Person	(Source numbers)
0	Start	(747–814)	Charlemagne the Great	(51, 144, 150, 1, 2)
1	Son	(778–840)	Louis I the Pious	(51, 144, 150, 216, 2)
2	Son	(795–855)	Lothaire I of Italy	(144, 150, 175, 216, 193)
3	Son	(835–869)	Lothair II of Lorraine	(51, 150, 216, 30, 107)
4	Daughter	(852–907)	Gisela of Lorraine	(150, 216, 107, 328, 1689)
5	Daughter	(868–917)	Ragnhildis von Friesland	(150, 216, 107)
6	Daughter	(896–968)	Matilda von Ringleheim	(51, 150, 2, 115, 107)
7	Daughter	(913–984)	Gerberge of Saxony	(51, 144, 150, 216, 193)
8	Daughter	(930–973)	Aubree of Lorraine	(216, 193, 2, 115, 107)
9	Son	(956–995)	Giselbert of Roucy	(221, 115, 107, 311, 1548)
10	Son	(988–1033)	Ebles I of Roucy	(193, 2, 115, 107, 259)
11	Daughter	(1014–1063)	Alice of Roucy	(216, 221, 2, 115, 107)
12	Daughter	(1032–1103)	Marguerite of Roucy	(166, 195, 2, 107, 259)
13	Daughter	(1072–)	Adeliza of Clermont	(2, 42, 227, 259, 293)
14	Son	(1104–1164)	John FitzGilbert le Mareschal	(2, 42, 227, 260, 293)

(continued)

Table C.6 (continued)

Gen	Relation	(Birth–Death)	Person	(Source numbers)
15	Son	(1146–1219)	William I le Mareschal	(175, 163, 216, 2, 37)
16	Daughter	(1198–1246)	Eve le Mareschal	(2, 37, 42, 80, 160)
17	Daughter	(1220–1255)	Eve de Braose	(173, 2, 37, 42, 259)
18	Daughter	(1250–1299)	Millicent de Cantelou	(146, 2, 42, 109, 67)
19	Son	(1271–1319)	Thomas de Greene	(109, 262, 956, 711, 22)
20	Son	(1292–1352)	Thomas de Greene	(109, 67, 234, 262, 293)
21	Son	(1310–1370)	Henry de Greene	(109, 67, 234, 262, 260)
22	Son	(1354–1399)	Henry II Greene	(155, 167, 67, 234, 260)
23	Daughter	(1380–1422)	Eleanor Greene	(155, 167, 260, 258, 42)
24	Son	(1415-)	John FitzWilliam	(155, 167, 146, 260, 258)
25	Son	(1455–1534)	William FitzWilliam	(146, 155, 144, 260, 333)
26	Daughter	(1504–1588)	Anne FitzWilliam	(155, 260, 258, 651, 711)
27	Daughter	(1528–1620)	Anne Cooke	(172, 260, 258, 334, 651)
28	Son	(1560–1626)	Francis Bacon	(184, 172, 259, 304, 258)

Table C.7 Lineage of maximal ascent from William Shakespeare to Charlemagne

Gen	Relation	(Birth–Death)	Person	(Source numbers)
0	Start	(747–814)	Charlemagne the Great	(51, 144, 150, 1, 2)
1	Son	(778–840)	Louis I the Pious	(51, 144, 150, 216, 2)
2	Son	(795–855)	Lothaire I of Italy	(144, 150, 175, 216, 193)
3	Son	(835–869)	Lothair II of Lorraine	(51, 150, 216, 30, 107)
4	Daughter	(852–907)	Gisela of Lorraine	(150, 216, 107, 328, 1689)
5	Daughter	(868–917)	Ragnhildis von Friesland	(150, 216, 107)
6	Daughter	(896–968)	Matilda von Ringleheim	(51, 150, 2, 115, 107)
7	Daughter	(913–984)	Gerberge of Saxony	(51, 144, 150, 216, 193)
8	Daughter	(930–973)	Aubree of Lorraine	(216, 193, 2, 115, 107)
9	Daughter	(950–1005)	Ermentrude de Roucy	(216, 2, 115, 107, 315)
10	Son	(990–1057)	Raynalt I of Burgundy	(51, 161, 216, 2, 95)
11	Son	(1024–1087)	William I of Burgundy	(144, 172, 216, 193, 2)
12	Daughter	(1062–1102)	Matilda of Burgundy	(216, 2, 115, 107, 315)

(continued)

Table C.7 (continued)

Gen	Relation	(Birth–Death)	Person	(Source numbers)
13	Daughter	(1080–1142)	Helie of Burgundy	(216, 2, 115, 107, 227)
14	Daughter	(1118–1174)	Ela of Ponthieu	(216, 2, 260, 293, 258)
15	Son	(1150–1196)	William FitzPatrick	(51, 161, 173, 2, 227)
16	Daughter	(1185–1261)	Ela of Salisbury	(51, 161, 173, 2, 42)
17	Daughter	(1220–)	Ida Longespee	(2, 42, 109, 259, 287)
18	Daughter	(1249–)	Ela FitzWalter	(144, 227, 259, 260, 287)
19	Daughter	(1266–1321)	Ida Odyngsells	(144, 260, 109, 259, 258)
20	Son	(1300–1335)	John II Clinton	(144, 259, 260, 287, 311)
21	Daughter	(1320–)	Ida de Clinton	(260)
22	Son	(1349–1387)	Baldwin Freville	(165, 259, 260, 293, 258)
23	Son	(1368–1416)	Baldwin Freville	(165, 2, 42, 109, 247)
24	Daughter	(1400–1493)	Margaret de Freville	(260, 258, 42)
25	Daughter	(1419–)	Joyce Willoughby	(260)
26	Daughter	(1469–1539)	Alice Bracebridge	(711)
27	Daughter	(1488–1548)	Margaret Arden	(711)
28	Daughter	(1512–1550)	Mary Agnes Webb	(711)
29	Daughter	(1538–1608)	Mary Arden	(507, 711)
30	Son	(1564–1616)	William Shakespeare	(304, 1318, 711, 1502, 507)

Table C.8 Lineage of maximal ascent from John Locke to Charlemagne

Gen	Relation	(Birth–Death)	Person	(Source numbers)
0	Start	(747–814)	Charlemagne the Great	(51, 144, 150, 1, 2)
1	Son	(778–840)	Louis I the Pious	(51, 144, 150, 216, 2)
2	Son	(795–855)	Lothaire I of Italy	(144, 150, 175, 216, 193)
3	Son	(835–869)	Lothair II of Lorraine	(51, 150, 216, 30, 107)
4	Daughter	(852–907)	Gisela of Lorraine	(150, 216, 107, 328, 1689)
5	Daughter	(868–917)	Ragnhildis von Friesland	(150, 216, 107)
6	Daughter	(896–968)	Matilda von Ringleheim	(51, 150, 2, 115, 107)
7	Daughter	(913–984)	Gerberge of Saxony	(51, 144, 150, 216, 193)

(continued)

Table C.8 (continued)

Gen	Relation	(Birth–Death)	Person	(Source numbers)
8	Daughter	(930–973)	Aubree of Lorraine	(216, 193, 2, 115, 107)
9	Daughter	(950–1005)	Ermentrude de Roucy	(216, 2, 115, 107, 315)
10	Son	(990–1057)	Raynalt I of Burgundy	(51, 161, 216, 2, 95)
11	Son	(1024–1087)	William I of Burgundy	(144, 172, 216, 193, 2)
12	Daughter	(1070–1135)	Gisele of Burgundy	(172, 216, 2, 115, 107)
13	Daughter	(1092–1154)	Adelaide de Savoy	(135, 172, 216, 224, 2)
14	Son	(1126–1183)	Peter I de Courtenay	(161, 163, 216, 221, 2)
15	Daughter	(1160–1218)	Alice de Courtenay	(216, 221, 2, 115, 107)
16	Daughter	(1188–1246)	Isabel Taillefer	(144, 195, 216, 2, 42)
17	Son	(1206–1272)	Henry III of England	(144, 171, 216, 2, 37)
18	Son	(1239–1307)	Edward I Plantagenet	(51, 161, 175, 171, 167)
19	Daughter	(1282–1316)	Elizabeth Plantagenet	(161, 184, 171, 216, 42)
20	Daughter	(1311–1391)	Margaret de Bohun	(161, 171, 42, 152, 259)
21	Son	(1341–1406)	Philip Courtenay	(161, 260, 258, 42, 1376)
22	Son	(1384–1406)	John Courtenay	(161, 260, 258, 566, 42)
23	Son	(1404–1463)	Philip Courtenay	(161, 260, 287, 258, 589)
24	Daughter	(1432–1493)	Elizabeth Courtenay	(258, 589, 711, 260)
25	Daughter	(1470–)	Philippe Audley	(589, 260, 711)
26	Son	(1495–1537)	James Hadley	(852)
27	Son	(1520–)	Christopher Hadley	(852)
28	Daughter	(1538–1607)	Margaret Hadley	(852)
29	Daughter	(1555–1638)	Margaret Strode	(956, 852)
30	Son	(1573–1630)	Edmund Keene	(956)
31	Daughter	(1597–1637)	Agnes Kenn	(956, 852)
32	Son	(1632–1704)	John Locke	(175, 247, 304, 1318, 1502)

Table C.9. Lineage of maximal ascent from Antoni van Leeuwenhoek to Charlemagne

Gen	Relation	(Birth–Death)	Person	(Source numbers)
0	Start	(747–814)	Charlemagne the Great	(51, 144, 150, 1, 2)
1	Son	(778–840)	Louis I the Pious	(51, 144, 150, 216, 2)
2	Son	(795–855)	Lothaire I of Italy	(144, 150, 175, 216, 193)
3	Son	(835–869)	Lothair II of Lorraine	(51, 150, 216, 30, 107)
4	Daughter	(852–907)	Gisela of Lorraine	(150, 216, 107, 328, 1689)
5	Daughter	(868–917)	Ragnhildis von Friesland	(150, 216, 107)
6	Daughter	(896–968)	Matilda von Ringleheim	(51, 150, 2, 115, 107)
7	Daughter	(913–984)	Gerberge of Saxony	(51, 144, 150, 216, 193)
8	Son	(953–991)	Charles of Lorraine	(216, 193, 150, 2, 30)
9	Daughter	(972–1019)	Adelheid of Lorraine	(150, 216, 52, 109, 107)
10	Son	(1000–1064)	Albert II of Namur	(193, 52, 107, 259, 311)
11	Son	(1030–1102)	Albert III of Namur	(150, 224, 221, 193, 52)
12	Son	(1072–1138)	Henry I de Namur	(224, 193, 227, 107)
13	Daughter	(1115–1169)	Mathilde de la Roche	(107, 1263, 1660, 115, 1137)
14	Son	(1155–1191)	Jacques I d'Avesnes	(221, 107, 311, 315, 1664)
15	Son	(1180–1244)	Bouchard d'Avesnes	(216, 2, 106, 115, 107)
16	Son	(1218–1257)	Jean I d'Avesnes	(216, 2, 106, 115, 107)
17	Son	(1247–1304)	John II d'Avesnes	(193, 162, 216, 138, 224)
18	Son	(1280–1335)	Willem de Cuser	(258, 1137)
19	Son	(1315–1407)	Coenraad Cuser	(258, 1137)
20	Daughter	(1340–)	Coensdr Cuser	(1137)
21	Son	(1370–1442)	Gerijt Van Spaemwoude	(1137)
22	Son	(1432–1509)	Ysbrant Van Spaemwoude	(1137)
23	Daughter	(1495–1528)	Clara Van Spaemwoude	(1137)
24	Daughter	(1525–1601)	Anna Jandr van der Mije	(1137)
25	Daughter	(1543–1627)	Neeltje van Hoogenhouck	(1137)
26	Son	(1560–1615)	Jacob van den Berch	(1137)
27	Daughter	(1595–1666)	Margriete Bel van den Berch	(1137, 566)
28	Son	(1632–1723)	Antoni van Leeuwenhoek	(1318, 1502, 1137, 304, 566)

Table C.10 Lineage of maximal ascent from Isaac Newton to Charlemagne

Gen	Relation	(Birth–Death)	Person	(Source numbers)
0	Start	(747–814)	Charlemagne the Great	(51, 144, 150, 1, 2)
1	Son	(778–840)	Louis I the Pious	(51, 144, 150, 216, 2)
2	Son	(795–855)	Lothaire I of Italy	(144, 150, 175, 216, 193)
3	Son	(835–869)	Lothair II of Lorraine	(51, 150, 216, 30, 107)
4	Daughter	(852–907)	Gisela of Lorraine	(150, 216, 107, 328, 1689)
5	Daughter	(868–917)	Ragnhildis von Friesland	(150, 216, 107)
6	Daughter	(896–968)	Matilda von Ringleheim	(51, 150, 2, 115, 107)
7	Daughter	(913–984)	Gerberge of Saxony	(51, 144, 150, 216, 193)
8	Daughter	(930–973)	Aubree of Lorraine	(216, 193, 2, 115, 107)
9	Daughter	(950–1005)	Ermentrude de Roucy	(216, 2, 115, 107, 315)
10	Son	(990–1057)	Raynalt I of Burgundy	(51, 161, 216, 2, 95)
11	Son	(1024–1087)	William I of Burgundy	(144, 172, 216, 193, 2)
12	Daughter	(1062–1102)	Matilda of Burgundy	(216, 2, 115, 107, 315)
13	Daughter	(1080–1142)	Helie of Burgundy	(216, 2, 115, 107, 227)
14	Daughter	(1118–1174)	Ela of Ponthieu	(216, 2, 260, 293, 258)
15	Daughter	(1137–1199)	Isabel de Warenne	(155, 216, 2, 73, 95)
16	Daughter	(1164–)	Isabel de Warenne	(260, 293)
17	Son	(1191–1225)	Hugh II Le Bigod	(2, 42, 80, 37, 260)
18	Daughter	(1212–)	Isabel Le Bigod	(165, 2, 42, 260, 287)
19	Daughter	(1240–1301)	Maud FitzJohn	(167, 172, 2, 42, 260)
20	Daughter	(1266–1306)	Isabel de Beauchamp	(51, 161, 165, 172, 216)
21	Daughter	(1293–1334)	Isabel le Despenser	(51, 161, 166, 109, 2)
22	Son	(1310–1347)	Hugh de Hastings	(161, 167, 260, 109)
23	Son	(1330–1369)	Hugh Hastings	(51, 109, 260)
24	Daughter	(1356–1384)	Joan Hastings	(162, 260)
25	Son	(1375–1401)	Robert de Morley	(175, 260)
26	Son	(1393–1435)	Thomas de Morley	(162, 175, 260, 258, 109)
27	Daughter	(1413–1471)	Anne de Morley	(258, 711, 260)
28	Daughter	(1437–1489)	Elizabeth Hastings	(258, 589, 711, 260)
29	Daughter	(1467–1492)	Margery Hildyard	(258, 711, 260)
30	Son	(1486–1540)	William Ayscough	(711, 258)
31	Son	(1509–1564)	Francis Ayscough	(711)
32	Son	(1540–1607)	Roger Ayscough	(711)
33	Son	(1585–)	James Ayscough	(711)
34	Daughter	(1623–1679)	Hannah Ayscough	(268, 711)
35	Son	(1642–1727)	Isaac Newton	(155, 268, 304, 1318, 1502)

Table C.11 Lineage of maximal ascent from Benjamin Franklin to Charlemagne

Gen	Relation	(Birth–Death)	Person	(Source numbers)
0	Start	(747–814)	Charlemagne the Great	(51, 144, 150, 1, 2)
1	Son	(778–840)	Louis I the Pious	(51, 144, 150, 216, 2)
2	Son	(795–855)	Lothaire I of Italy	(144, 150, 175, 216, 193)
3	Son	(835–869)	Lothair II of Lorraine	(51, 150, 216, 30, 107)
4	Daughter	(852–907)	Gisela of Lorraine	(150, 216, 107, 328, 1689)
5	Daughter	(868–917)	Ragnhildis von Friesland	(150, 216, 107)
6	Daughter	(896–968)	Matilda von Ringleheim	(51, 150, 2, 115, 107)
7	Daughter	(913–984)	Gerberge of Saxony	(51, 144, 150, 216, 193)
8	Daughter	(930–973)	Aubree of Lorraine	(216, 193, 2, 115, 107)
9	Daughter	(950–1005)	Ermentrude de Roucy	(216, 2, 115, 107, 315)
10	Son	(990–1057)	Raynalt I of Burgundy	(51, 161, 216, 2, 95)
11	Son	(1024–1087)	William I of Burgundy	(144, 172, 216, 193, 2)
12	Daughter	(1062–1102)	Matilda of Burgundy	(216, 2, 115, 107, 315)
13	Daughter	(1080–1142)	Helie of Burgundy	(216, 2, 115, 107, 227)
14	Daughter	(1118–1174)	Ela of Ponthieu	(216, 2, 260, 293, 258)
15	Daughter	(1137–1199)	Isabel de Warenne	(155, 216, 2, 73, 95)
16	Daughter	(1164–)	Isabel de Warenne	(260, 293)
17	Son	(1191–1225)	Hugh II Le Bigod	(2, 42, 80, 37, 260)
18	Daughter	(1212–)	Isabel Le Bigod	(165, 2, 42, 260, 287)
19	Daughter	(1240–1301)	Maud FitzJohn	(167, 172, 2, 42, 260)
20	Daughter	(1266–1306)	Isabel de Beauchamp	(51, 161, 165, 172, 216)
21	Daughter	(1282–1322)	Maud de Chaworth	(216, 2, 42, 149, 260)
22	Daughter	(1312–1349)	Joan of Lancaster	(158, 171, 216, 2, 42)
23	Son	(1340–1368)	John III de Mowbray	(51, 158, 162, 2, 42)
24	Daughter	(1357–1401)	Margaret de Mowbray	(258, 260)
25	Son	(1380–1444)	Walter Lucy	(173, 260, 258)
26	Daughter	(1405–1460)	Eleanor Lucy	(173, 260, 258)
27	Daughter	(1427–1498)	Elizabeth Hopton	(173, 260, 258)
28	Daughter	(1463–1525)	Joan Stanley	(711, 260)
29	Daughter	(1500–1583)	Elizabeth Warburton	(711)
30	Daughter	(1523–1583)	Agnes Holmes	(1146, 711)
31	Daughter	(1556–1655)	Margaret Lawrence	(711)
32	Daughter	(1574–1658)	Alice Elmy	(711, 247)

(continued)

Table C.11 (continued)

Gen	Relation	(Birth–Death)	Person	(Source numbers)
33	Daughter	(1595–1664)	Meribah Gibbs	(711, 247)
34	Son	(1618–1690)	Peter Folger	(763, 247)
35	Daughter	(1667–1752)	Abiah Folger	(247, 1376)
36	Son	(1706–1790)	Benjamin Franklin	(304, 1318, 711, 1502, 247)

Table C.12. Lineage of maximal ascent from Adam Smith to Charlemagne

Gen	Relation	(Birth–Death)	Person	(Source numbers)
0	Start	(747–814)	Charlemagne the Great	(51, 144, 150, 1, 2)
1	Son	(778–840)	Louis I the Pious	(51, 144, 150, 216, 2)
2	Son	(795–855)	Lothaire I of Italy	(144, 150, 175, 216, 193)
3	Son	(835–869)	Lothair II of Lorraine	(51, 150, 216, 30, 107)
4	Daughter	(852–907)	Gisela of Lorraine	(150, 216, 107, 328, 1689)
5	Daughter	(868–917)	Ragnhildis von Friesland	(150, 216, 107)
6	Daughter	(896–968)	Matilda von Ringleheim	(51, 150, 2, 115, 107)
7	Daughter	(913–984)	Gerberge of Saxony	(51, 144, 150, 216, 193)
8	Daughter	(930–973)	Aubree of Lorraine	(216, 193, 2, 115, 107)
9	Daughter	(958–1009)	Berthilde de Marle	(1680, 1154)
10	Daughter	(975–1031)	Jehanne de Coucy	(1689, 1154)
11	Daughter	(995–1058)	Pavie de Ham	(150, 2, 6, 115, 107)
12	Son	(1032–1080)	Herbert IV de Vermandoise	(150, 2, 6, 115, 311)
13	Daughter	(1062–1120)	Adele de Vermandoise	(150, 216, 2, 84, 115)
14	Daughter	(1081–1147)	Isabel de Vermandoise	(155, 175, 172, 173, 216)
15	Daughter	(1104–1178)	Ada de Warenne	(216, 2, 52, 66, 95)
16	Son	(1144–1219)	David of Huntingdon	(155, 158, 161, 175, 164)
17	Daughter	(1206–1251)	Isabella of Huntingdon	(158, 166, 216, 9, 42)
18	Son	(1224–1295)	Robert Bruce	(158, 166, 216, 9, 42)
19	Son	(1243–1304)	Robert David Bruce	(158, 166, 172, 170, 216)
20	Son	(1274–1329)	Robert I Bruce	(158, 164, 175, 179, 173)
21	Daughter	(1296–1316)	Marjorie Bruce	(158, 179, 216, 9, 42)
22	Son	(1316–1390)	Robert II Stewart	(144, 158, 175, 172, 162)
23	Son	(1349–1420)	Robert Stewart	(144, 161, 175, 164, 179)

(continued)

Table C.12. (continued)

Gen	Relation	(Birth–Death)	Person	(Source numbers)
24	Daughter	(1372–)	Isabella Stewart	(172, 216, 287, 956)
25	Daughter	(1415–)	Christian Halyburton	(167, 172, 287, 42)
26	Son	(1436–1473)	Andrew Leslie	(172, 589)
27	Son	(1460–1513)	William Leslie	(172, 589, 956)
28	Son	(1485–1558)	George Leslie	(164, 167, 166, 179, 51)
29	Daughter	(1539–1594)	Agnes Leslie	(179, 167, 244, 502, 589)
30	Son	(1568–1609)	George Douglas	(179, 226, 589, 502)
31	Son	(1600–1620)	Archibald Douglas	(226, 501, 502)
32	Son	(1620–)	John Douglas	(504, 502)
33	Son	(1653–1706)	Robert Douglas	(226, 503, 502)
34	Daughter	(1694–1784)	Margaret Douglas	(502)
35	Son	(1723–1790)	Adam Smith	(304, 1318, 1502, 502, 493)

Table C.13 Lineage of maximal ascent from Robert Clive to Charlemagne

Gen	Relation	(Birth–Death)	Person	(Source numbers)
0	Start	(747–814)	Charlemagne the Great	(51, 144, 150, 1, 2)
1	Son	(778–840)	Louis I the Pious	(51, 144, 150, 216, 2)
2	Son	(795–855)	Lothaire I of Italy	(144, 150, 175, 216, 193)
3	Son	(835–869)	Lothair II of Lorraine	(51, 150, 216, 30, 107)
4	Daughter	(852–907)	Gisela of Lorraine	(150, 216, 107, 328, 1689)
5	Daughter	(868–917)	Ragnhildis von Friesland	(150, 216, 107)
6	Daughter	(896–968)	Matilda von Ringleheim	(51, 150, 2, 115, 107)
7	Daughter	(913–984)	Gerberge of Saxony	(51, 144, 150, 216, 193)
8	Daughter	(930–973)	Aubree of Lorraine	(216, 193, 2, 115, 107)
9	Daughter	(950–1005)	Ermentrude de Roucy	(216, 2, 115, 107, 315)
10	Son	(990–1057)	Raynalt I of Burgundy	(51, 161, 216, 2, 95)
11	Son	(1024–1087)	William I of Burgundy	(144, 172, 216, 193, 2)
12	Daughter	(1070–1135)	Gisele of Burgundy	(172, 216, 2, 115, 107)
13	Daughter	(1092–1154)	Adelaide de Savoy	(135, 172, 216, 224, 2)
14	Son	(1126–1183)	Peter I de Courtenay	(161, 163, 216, 221, 2)
15	Daughter	(1160–1218)	Alice de Courtenay	(216, 221, 2, 115, 107)

(continued)

Table C.13 (continued)

Gen	Relation	(Birth–Death)	Person	(Source numbers)
16	Daughter	(1188–1246)	Isabel Taillefer	(144, 195, 216, 2, 42)
17	Son	(1206–1272)	Henry III of England	(144, 171, 216, 2, 37)
18	Son	(1239–1307)	Edward I Plantagenet	(51, 161, 175, 171, 167)
19	Daughter	(1272–1307)	Joan of Acre	(51, 161, 175, 167, 216)
20	Daughter	(1293–1342)	Margaret de Clare	(2, 37, 42, 80, 95)
21	Daughter	(1312–1357)	Amy de Gaveston	(258, 109, 1376)
22	Daughter	(1340–1412)	Alice de Driby	(259, 260, 258, 956, 109)
23	Son	(1365–1445)	William Malory	(259, 258, 109, 1376)
24	Daughter	(1387–1439)	Margaret Malory	(259, 260, 258, 109, 1376)
25	Son	(1415–1467)	Roger Corbet	(173, 155, 260, 258)
26	Son	(1451–1493)	Richard Corbet	(161, 158, 51, 153, 260)
27	Daughter	(1478–)	Margaret Corbet	(158, 260)
28	Son	(1495–1573)	Richard Clive	(158)
29	Son	(1540–1591)	George Clive	(158)
30	Son	(1585–)	Ambrose Clive	(158)
31	Son	(1615–)	Robert Clive	(158)
32	Son	(1645–)	George Clive	(158)
33	Son	(1675–)	Robert Clive	(158)
34	Son	(1704–)	Richard Clive	(158)
35	Son	(1726–1774)	Robert Clive	(158, 304, 1318)

Table C.14 Lineage of maximal ascent from James Watt to Charlemagne

Gen	Relation	(Birth–Death)	Person	(Source numbers)
0	Start	(747–814)	Charlemagne the Great	(51, 144, 150, 1, 2)
1	Son	(778–840)	Louis I the Pious	(51, 144, 150, 216, 2)
2	Daughter	(798–860)	Hildegarde Rotrude Adeltrude of France	(150, 938, 711, 107, 30)
3	Son	(815–857)	Erispoe II de Vannes	(1, 52, 95, 107, 1689)
4	Daughter	(830–880)	Lotitia de Poher	(328, 1154, 1689, 107)
5	Son	(860–888)	Berenger Judicael I of Rennes	(216, 328, 1689, 107)
6	Daughter	(883–925)	Papie de Bayeux	(216, 2, 95, 115, 107)

(continued)

Table C.14 (continued)

Gen	Relation	(Birth–Death)	Person	(Source numbers)
7	Son	(900–942)	William I Long Sword	(150, 175, 216, 2, 32)
8	Son	(933–996)	Richard I of Normandy	(175, 184, 171, 216, 2)
9	Son	(970–1027)	Richard II of Normandy	(51, 150, 161, 216, 2)
10	Son	(1002–1035)	Robert I of Normandy	(51, 158, 161, 164, 166)
11	Son	(1028–1087)	William I the Conqueror	(158, 164, 171, 169, 166)
12	Son	(1068–1135)	Henry I Beauclerc	(144, 150, 155, 175, 163)
13	Son	(1090–1147)	Robert de Caen	(172, 166, 2, 73, 95)
14	Daughter	(1110–1189)	Maud de Caen	(166, 2, 42, 240, 259)
15	Son	(1147–1181)	Hugh of Kevelioc	(144, 158, 161, 163, 169)
16	Daughter	(1171–1233)	Maud of Chester	(158, 161, 166, 216, 2)
17	Daughter	(1206–1251)	Isabella of Huntingdon	(158, 166, 216, 9, 42)
18	Son	(1224–1295)	Robert Bruce	(158, 166, 216, 9, 42)
19	Son	(1243–1304)	Robert David Bruce	(158, 166, 172, 170, 216)
20	Son	(1274–1329)	Robert I Bruce	(158, 164, 175, 179, 173)
21	Daughter	(1296–1316)	Marjorie Bruce	(158, 179, 216, 9, 42)
22	Son	(1316–1390)	Robert II Stewart	(144, 158, 175, 172, 162)
23	Daughter	(1351–)	Elizabeth Stewart	(175, 172, 216, 287, 203)
24	Son	(1374–1437)	William Hay	(167, 175, 1162)
25	Daughter	(1394–)	Jean Hay	(1162)
26	Son	(1415–1506)	William Muirhead	(1162)
27	Son	(1443–1513)	John Muirhead	(1162)
28	Son	(1466–1492)	John Muirhead	(711, 1162)
29	Son	(1486–)	James Muirhead	(711, 1162)
30	Son	(1510–1568)	James Muirhead	(711)
31	Son	(1531–1622)	James Muirhead	(711, 1376)
32	Son	(1573–1633)	James Muirhead	(1163, 711)
33	Son	(1595–)	Alexander Muirhead	(711)
34	Son	(1622–)	John Muirhead	(1163)
35	Son	(1662–1727)	Robert Muirhead	(1163)
36	Daughter	(1703–1755)	Agnes Muirhead	(711, 1163)
37	Son	(1736–1819)	James Watt	(566, 304, 1318, 1502, 1163)

Table C.15 Lineage of maximal ascent from Eli Whitney to Charlemagne

Gen	Relation	(Birth–Death)	Person	(Source numbers)
0	Start	(747–814)	Charlemagne the Great	(51, 144, 150, 1, 2)
1	Son	(778–840)	Louis I the Pious	(51, 144, 150, 216, 2)
2	Son	(795–855)	Lothaire I of Italy	(144, 150, 175, 216, 193)
3	Son	(835–869)	Lothair II of Lorraine	(51, 150, 216, 30, 107)
4	Daughter	(852–907)	Gisela of Lorraine	(150, 216, 107, 328, 1689)
5	Daughter	(868–917)	Ragnhildis von Friesland	(150, 216, 107)
6	Daughter	(896–968)	Matilda von Ringleheim	(51, 150, 2, 115, 107)
7	Daughter	(913–984)	Gerberge of Saxony	(51, 144, 150, 216, 193)
8	Daughter	(930–973)	Aubree of Lorraine	(216, 193, 2, 115, 107)
9	Daughter	(950–1005)	Ermentrude de Roucy	(216, 2, 115, 107, 315)
10	Son	(990–1057)	Raynalt I of Burgundy	(51, 161, 216, 2, 95)
11	Son	(1024–1087)	William I of Burgundy	(144, 172, 216, 193, 2)
12	Daughter	(1062–1102)	Matilda of Burgundy	(216, 2, 115, 107, 315)
13	Daughter	(1080–1142)	Helie of Burgundy	(216, 2, 115, 107, 227)
14	Daughter	(1118–1174)	Ela of Ponthieu	(216, 2, 260, 293, 258)
15	Daughter	(1137–1199)	Isabel de Warenne	(155, 216, 2, 73, 95)
16	Daughter	(1164–)	Isabel de Warenne	(260, 293)
17	Son	(1191–1225)	Hugh II Le Bigod	(2, 42, 80, 37, 260)
18	Daughter	(1212–)	Isabel Le Bigod	(165, 2, 42, 260, 287)
19	Daughter	(1240–1301)	Maud FitzJohn	(167, 172, 2, 42, 260)
20	Daughter	(1266–1306)	Isabel de Beauchamp	(51, 161, 165, 172, 216)
21	Daughter	(1293–1334)	Isabel le Despenser	(51, 161, 166, 109, 2)
22	Son	(1310–1347)	Hugh de Hastings	(161, 167, 260, 109)
23	Son	(1330–1369)	Hugh Hastings	(51, 109, 260)
24	Daughter	(1356–1384)	Joan Hastings	(162, 260)
25	Son	(1375–1401)	Robert de Morley	(175, 260)
26	Son	(1395–)	William Morley	(161, 6)
27	Daughter	(1416–1503)	Margary Morley	(161, 6)

(continued)

Table C.15 (continued)

Gen	Relation	(Birth–Death)	Person	(Source numbers)
28	Son	(1437–1474)	William Yelverton	(161, 260, 1589, 6)
29	Daughter	(1455–)	Margaret Yelverton	(161, 260, 287, 1589, 42)
30	Son	(1473–1516)	Henry Palgrave	(234, 240, 259, 260, 287)
31	Son	(1507–1545)	Thomas Palgrave	(234, 240, 259, 260, 287)
32	Son	(1540–1623)	Edward Palgrave	(234, 240, 260, 287, 258)
33	Son	(1585–1651)	Richard Palgrave	(234, 240, 259, 260, 287)
34	Daughter	(1619–1695)	Mary Palgrave	(234, 240, 259, 260, 287)
35	Son	(1647–1710)	Benjamin Wellington	(234, 240, 259, 307, 258)
36	Daughter	(1673–1729)	Elizabeth Wellington	(234, 240, 259, 258, 109)
37	Son	(1701–)	Benjamin Fay	(240)
38	Daughter	(1740–1777)	Elizabeth Fay	(247, 1146, 240)
39	Son	(1765–1825)	Eli Whitney	(247, 304, 1318, 240, 1146)

Table C.16 Lineage of maximal ascent from Robert Fulton to Charlemagne

Gen	Relation	(Birth–Death)	Person	(Source numbers)
0	Start	(747–814)	Charlemagne the Great	(51, 144, 150, 1, 2)
1	Son	(778–840)	Louis I the Pious	(51, 144, 150, 216, 2)
2	Son	(795–855)	Lothaire I of Italy	(144, 150, 175, 216, 193)
3	Son	(835–869)	Lothair II of Lorraine	(51, 150, 216, 30, 107)
4	Daughter	(852–907)	Gisela of Lorraine	(150, 216, 107, 328, 1689)
5	Daughter	(868–917)	Ragnhildis von Friesland	(150, 216, 107)
6	Daughter	(896–968)	Matilda von Ringleheim	(51, 150, 2, 115, 107)
7	Daughter	(913–984)	Gerberge of Saxony	(51, 144, 150, 216, 193)
8	Daughter	(930–973)	Aubree of Lorraine	(216, 193, 2, 115, 107)
9	Daughter	(950–1005)	Ermentrude de Roucy	(216, 2, 115, 107, 315)
10	Son	(990–1057)	Raynalt I of Burgundy	(51, 161, 216, 2, 95)
11	Son	(1024–1087)	William I of Burgundy	(144, 172, 216, 193, 2)
12	Daughter	(1070–1135)	Gisele of Burgundy	(172, 216, 2, 115, 107)
13	Daughter	(1092–1154)	Adelaide de Savoy	(135, 172, 216, 224, 2)
14	Son	(1126–1183)	Peter I de Courtenay	(161, 163, 216, 221, 2)
15	Daughter	(1160–1218)	Alice de Courtenay	(216, 221, 2, 115, 107)

(continued)

Table C.16 (continued)

Gen	Relation	(Birth–Death)	Person	(Source numbers)
16	Daughter	(1188–1246)	Isabel Taillefer	(144, 195, 216, 2, 42)
17	Son	(1223–1296)	William de Valence	(146, 161, 171, 195, 37)
18	Daughter	(1273–1326)	Joan de Valence	(164, 195, 37, 42, 80)
19	Daughter	(1299–1372)	Elizabeth Comyn	(146, 37, 42, 80, 260)
20	Son	(1332–1387)	Gilbert Talbot	(146, 184, 163, 37, 42)
21	Son	(1361–1396)	Richard Talbot	(146, 161, 37, 42, 80)
22	Daughter	(1388–)	Alice Talbot	(146, 260)
23	Daughter	(1411–)	Wenlian Barry	(144)
24	Daughter	(1434–)	Maud Vaughan	(144)
25	Son	(1464–1541)	David Sicelt	(144)
26	Son	(1500–1552)	Richard Cecil	(144, 260, 287, 42)
27	Daughter	(1545–1611)	Margaret Cecil	(144, 259, 258, 711)
28	Son	(1588–1648)	Henry Smith	(247, 711)
29	Son	(1619–1670)	Peregrine Smith	(711)
30	Son	(1635–1680)	William Smith	(711)
31	Son	(1660–)	John Smith	(711)
32	Son	(1704–1767)	Joseph Smith	(711)
33	Daughter	(1734–1780)	Mary Smith	(711)
34	Son	(1765–1815)	Robert Fulton	(493, 304, 1318, 258, 711)

Table C.17 Lineage of maximal ascent from Thomas Robert Malthus to Charlemagne

Gen	Relation	(Birth–Death)	Person	(Source numbers)
0	Start	(747–814)	Charlemagne the Great	(51, 144, 150, 1, 2)
1	Son	(778–840)	Louis I the Pious	(51, 144, 150, 216, 2)
2	Son	(795–855)	Lothaire I of Italy	(144, 150, 175, 216, 193)
3	Son	(835–869)	Lothair II of Lorraine	(51, 150, 216, 30, 107)
4	Daughter	(852–907)	Gisela of Lorraine	(150, 216, 107, 328, 1689)
5	Daughter	(868–917)	Ragnhildis von Friesland	(150, 216, 107)
6	Daughter	(896–968)	Matilda von Ringleheim	(51, 150, 2, 115, 107)
7	Daughter	(913–984)	Gerberge of Saxony	(51, 144, 150, 216, 193)

(continued)

Table C.17 (continued)

Gen	Relation	(Birth–Death)	Person	(Source numbers)
8	Daughter	(930–973)	Aubree of Lorraine	(216, 193, 2, 115, 107)
9	Son	(956–995)	Giselbert of Roucy	(221, 115, 107, 311, 1548)
10	Son	(988–1033)	Ebles I of Roucy	(193, 2, 115, 107, 259)
11	Daughter	(1014–1063)	Alice of Roucy	(216, 221, 2, 115, 107)
12	Daughter	(1032–1103)	Marguerite of Roucy	(166, 195, 2, 107, 259)
13	Daughter	(1072–)	Adeliza of Clermont	(2, 42, 227, 259, 293)
14	Daughter	(1092–1163)	Alice FitzGilbert de Clare	(2, 42, 259, 258, 1146)
15	Daughter	(1109–1166)	Rohais de Vere	(167, 2, 42, 109)
16	Daughter	(1138–)	Maud de Mandeville	(167, 2, 42, 227)
17	Son	(1156–1213)	Geoffrey III FitzPiers	(175, 167, 184, 2, 42)
18	Son	(1209–1258)	John FitzGeoffrey	(167, 165, 184, 172, 2)
19	Daughter	(1237–)	Isabel FitzJohn	(167, 2, 42, 287, 714)
20	Daughter	(1254–1292)	Isabel de Vipont	(51, 161, 175, 2, 42)
21	Son	(1274–1314)	Robert I de Clifford	(51, 144, 161, 175, 164)
22	Daughter	(1298–1365)	Idoine de Clifford	(144, 164, 2, 42, 260)
23	Daughter	(1320–1361)	Eleanor Percy	(144, 260, 109)
24	Daughter	(1357–1401)	Alice FitzWalter	(171, 260, 109)
25	Son	(1385–1417)	Richard de Vere	(171, 184, 260, 258, 109)
26	Son	(1408–1461)	John de Vere	(51, 184, 168, 260, 287)
27	Daughter	(1433–1507)	Jane de Vere	(270, 287, 711)
28	Daughter	(1462–)	Margaret Norreys	(711, 258)
29	Son	(1500–1558)	Thomas Bullock	(711, 258)
30	Daughter	(1534–1558)	Margaret Bullock	(711)
31	Son	(1557–1646)	Robert Malthus	(711)
32	Son	(1605–)	Robert Malthus	(711)
33	Son	(1651–1717)	Daniel Malthus	(711)
34	Son	(1685–1757)	Sydenham Malthus	(711)
35	Son	(1730–1800)	Daniel Malthus	(711)
36	Son	(1766–1834)	Thomas Robert Malthus	(304, 1318, 1502, 711)

Table C.18 Lineage of maximal ascent from Napoleon I Buonaparte to Charlemagne

Gen	Relation	(Birth–Death)	Person	(Source numbers)
0	Start	(747–814)	Charlemagne the Great	(51, 144, 150, 1, 2)
1	Son	(778–840)	Louis I the Pious	(51, 144, 150, 216, 2)
2	Son	(795–855)	Lothaire I of Italy	(144, 150, 175, 216, 193)
3	Son	(835–869)	Lothair II of Lorraine	(51, 150, 216, 30, 107)
4	Daughter	(852–907)	Gisela of Lorraine	(150, 216, 107, 328, 1689)
5	Daughter	(868–917)	Ragnhildis von Friesland	(150, 216, 107)
6	Daughter	(896–968)	Matilda von Ringleheim	(51, 150, 2, 115, 107)
7	Daughter	(913–984)	Gerberge of Saxony	(51, 144, 150, 216, 193)
8	Son	(953–991)	Charles of Lorraine	(216, 193, 150, 2, 30)
9	Daughter	(974–1018)	Gerberga of Lorraine	(216, 193, 150, 2, 52)
10	Son	(992–1062)	Lambert II of Brabant	(144, 150, 193, 2, 52)
11	Son	(1023–1077)	Reginald II of Brabant	(193, 1689, 2)
12	Daughter	(1050–1112)	Mahaud of Louvain	(328, 1689, 107)
13	Son	(1080–1135)	Henry I of Chatillon	(221, 311, 328, 1680, 107)
14	Son	(1116–1187)	Renaud I of Chatillon en Bazois	(3, 150, 216, 2, 35)
15	Daughter	(1162–1235)	Alecia of Antioch	(150, 102)
16	Son	(1205–1264)	Azzo VII d'Este	(150, 106, 342, 343, 102)
17	Daughter	(1232–)	Cubitosa d'Este	(150, 343, 102)
18	Son	(1260–1289)	Gabriele Malaspina	(343)
19	Son	(1289–)	Isnardo II Malaspina	(1393, 343)
20	Son	(1336–1418)	Niccolo Malaspina	(1393, 1407, 343)
21	Daughter	(1390–)	Apollonia Malaspina	(1407)
22	Son	(1425–1501)	Giovanni Buonaparte	(1407)
23	Son	(1460–1540)	Francesco II Mauro di Sarzana Buonaparte	(1407)
24	Son	(1491–1589)	Gabriele Buonaparte	(1407)
25	Son	(1520–1617)	Geronimo Buonaparte	(1407)
26	Son	(1577–1633)	Francesco Buonaparte	(1407)
27	Son	(1603–1642)	Sebastiano Buonaparte	(1407)
28	Son	(1637–1692)	Carlo Maria Buonaparte	(1407)
29	Son	(1663–1703)	Giuseppe Maria Buonaparte	(1407)
30	Son	(1683–1721)	Sebastiano Nicolo Buonaparte	(1407)
31	Son	(1713–1763)	Giuseppe Maria Buonaparte	(1407)
32	Son	(1746–1785)	Charles Marie Buonaparte	(216, 76, 113, 589, 52)

(continued)

Table C.18 (continued)

Gen	Relation	(Birth–Death)	Person	(Source numbers)
33	Son	(1769–1821)	Napoleon I Buonaparte	(78, 150, 154, 216, 76)

Table C.19 Lineage of maximal ascent from Simon Jose de Bolivar to Charlemagne

Gen	Relation	(Birth–Death)	Person	(Source numbers)
0	Start	(747–814)	Charlemagne the Great	(51, 144, 150, 1, 2)
1	Son	(778–840)	Louis I the Pious	(51, 144, 150, 216, 2)
2	Son	(795–855)	Lothaire I of Italy	(144, 150, 175, 216, 193)
3	Son	(835–869)	Lothair II of Lorraine	(51, 150, 216, 30, 107)
4	Daughter	(852–907)	Gisela of Lorraine	(150, 216, 107, 328, 1689)
5	Daughter	(868–917)	Ragnhildis von Friesland	(150, 216, 107)
6	Daughter	(896–968)	Matilda von Ringleheim	(51, 150, 2, 115, 107)
7	Daughter	(913–984)	Gerberge of Saxony	(51, 144, 150, 216, 193)
8	Daughter	(930–973)	Aubree of Lorraine	(216, 193, 2, 115, 107)
9	Daughter	(950–1005)	Ermentrude de Roucy	(216, 2, 115, 107, 315)
10	Son	(990–1057)	Raynalt I of Burgundy	(51, 161, 216, 2, 95)
11	Son	(1024–1087)	William I of Burgundy	(144, 172, 216, 193, 2)
12	Daughter	(1070–1135)	Gisele of Burgundy	(172, 216, 2, 115, 107)
13	Son	(1090–1148)	Amadeus III of Savoy	(216, 2, 96, 106, 113)
14	Daughter	(1125–1157)	Maud of Savoy	(216, 2, 108, 113, 115)
15	Daughter	(1150–1188)	Urraca of Portugal	(216, 2, 104, 108, 113)
16	Son	(1171–1230)	Alfonso Ferdinandez IX of Leon	(216, 2, 46, 49, 66)
17	Son	(1201–1252)	Ferdinand Alphonsez of Castile and Leon	(150, 166, 216, 2, 66)
18	Son	(1221–1284)	Alfonso X of Castile and Leon	(216, 2, 104, 106, 113)
19	Son	(1258–1295)	Sancho IV of Castile and Leon	(216, 104, 106, 113, 108)
20	Daughter	(1282–)	Teresa Sanchez of Castile and Leon	(711, 260)
21	Son	(1300–1354)	Fernan Rodriguez de Villalobos	(711)
22	Daughter	(1325–)	Maria Fernandez de Villalobos	(711)

(continued)

Table C.19 (continued)

Gen	Relation	(Birth–Death)	Person	(Source numbers)
23	Daughter	(1360–)	Maria Beatriz Osorio de Villalobos	(711)
24	Son	(1395–)	Diego Hernandez de Escobar	(711, 415)
25	Son	(1425–)	Alonso de Caceres y Escobar	(711, 415)
26	Son	(1455–)	Juan de Rojas y Escobar	(415)
27	Son	(1479–)	Lazaro Vasquez de Rojas	(415)
28	Daughter	(1503–1561)	Ana de Rojas	(415)
29	Daughter	(1541–)	Ana de Rojas	(415)
30	Daughter	(1560–)	Germana Diaz de Rojas	(415)
31	Daughter	(1580–)	Francisca Vasquez de Escobedo	(415)
32	Daughter	(1603–1646)	Maria Rodriguez Santos y Vasquez	(415)
33	Daughter	(1631–)	Juana de Rivilla y Puerta	(415)
34	Daughter	(1659–1723)	Josefa Fernandez de Araujo	(415)
35	Son	(1689–1742)	Mateo Jose Blanco y Fernandez de Araujo	(415)
36	Daughter	(1732–)	Francisca Blanco y Herrera	(415)
37	Daughter	(1759–1792)	Maria de la Concepcion Palacios y Blanco	(415, 414)
38	Son	(1783–1830)	Simon Jose Antonia de la de Bolivar	(414, 304, 1318, 1502, 415)

Table C.20 Lineage of maximal ascent from Victor Hugo to Charlemagne

Gen	Relation	(Birth–Death)	Person	(Source numbers)
0	Start	(747–814)	Charlemagne the Great	(51, 144, 150, 1, 2)
1	Son	(778–840)	Louis I the Pious	(51, 144, 150, 216, 2)
2	Son	(795–855)	Lothaire I of Italy	(144, 150, 175, 216, 193)
3	Son	(835–869)	Lothair II of Lorraine	(51, 150, 216, 30, 107)
4	Daughter	(852–907)	Gisela of Lorraine	(150, 216, 107, 328, 1689)
5	Daughter	(868–917)	Ragnhildis von Friesland	(150, 216, 107)
6	Daughter	(896–968)	Matilda von Ringleheim	(51, 150, 2, 115, 107)

(continued)

Table C.20 (continued)

Gen	Relation	(Birth–Death)	Person	(Source numbers)
7	Daughter	(913–984)	Gerberge of Saxony	(51, 144, 150, 216, 193)
8	Daughter	(930–973)	Aubree of Lorraine	(216, 193, 2, 115, 107)
9	Daughter	(950–1005)	Ermentrude de Roucy	(216, 2, 115, 107, 315)
10	Son	(990–1057)	Raynalt I of Burgundy	(51, 161, 216, 2, 95)
11	Son	(1024–1087)	William I of Burgundy	(144, 172, 216, 193, 2)
12	Daughter	(1070–1135)	Gisele of Burgundy	(172, 216, 2, 115, 107)
13	Daughter	(1092–1154)	Adelaide de Savoy	(135, 172, 216, 224, 2)
14	Son	(1123–1188)	Robert I of Dreux	(135, 172, 206, 216, 221)
15	Daughter	(1145–1217)	Alix of Dreux	(221, 195, 258, 936, 1154)
16	Son	(1167–1219)	Gaucher III de Chatillon-sur-Marne	(221, 711)
17	Daughter	(1202–1240)	Eustachie de Chatillon	(711)
18	Son	(1235–1284)	Hellin III de Wavrin	(1191, 711)
19	Son	(1250–1308)	Robert I de Wavrin	(711)
20	Daughter	(1270–)	Jeanne de Wavrin	(711)
21	Son	(1300–1355)	Jean III de Malet	(260, 1155)
22	Daughter	(1335–)	Isabelle Malet	(1155)
23	Daughter	(1370–)	Marie de Creully	(1155)
24	Daughter	(1405–)	Thomasse de Vierville	(1155)
25	Son	(1440–1513)	Philippe de Vassy	(1155)
26	Daughter	(1480–)	Renee de Vassy	(1155)
27	Son	(1509–1595)	Christophe du Breil	(711, 1155)
28	Son	(1530–1577)	Thomas du Breil	(1155)
29	Daughter	(1548–1610)	Perrine du Breil	(1155)
30	Son	(1575–)	Jean du Rocher	(1155)
31	Daughter	(1597–)	Bertrande du Rocher	(1155)
32	Son	(1618–)	Sulpice Handorin	(1155)
33	Daughter	(1648–)	Francois Handorin	(1155)
34	Daughter	(1691–)	Francoise Louvigne	(1155)

(continued)

Table C.20 (continued)

Gen	Relation	(Birth–Death)	Person	(Source numbers)
35	Son	(1731–1783)	Jean Francois Trebuchet	(1155)
36	Daughter	(1772–1821)	Sophie Trebuchet	(1155)
37	Son	(1802–1885)	Victor Hugo	(304, 1318, 1155, 566)

Table C.21 Lineage of maximal ascent from Charles Robert Darwin to Charlemagne

Gen	Relation	(Birth–Death)	Person	(Source numbers)
0	Start	(747–814)	Charlemagne the Great	(51, 144, 150, 1, 2)
1	Son	(778–840)	Louis I the Pious	(51, 144, 150, 216, 2)
2	Son	(795–855)	Lothaire I of Italy	(144, 150, 175, 216, 193)
3	Son	(835–869)	Lothair II of Lorraine	(51, 150, 216, 30, 107)
4	Daughter	(852–907)	Gisela of Lorraine	(150, 216, 107, 328, 1689)
5	Daughter	(868–917)	Ragnhildis von Friesland	(150, 216, 107)
6	Daughter	(896–968)	Matilda von Ringleheim	(51, 150, 2, 115, 107)
7	Daughter	(913–984)	Gerberge of Saxony	(51, 144, 150, 216, 193)
8	Daughter	(930–973)	Aubree of Lorraine	(216, 193, 2, 115, 107)
9	Daughter	(950–1005)	Ermentrude de Roucy	(216, 2, 115, 107, 315)
10	Son	(990–1057)	Raynalt I of Burgundy	(51, 161, 216, 2, 95)
11	Son	(1024–1087)	William I of Burgundy	(144, 172, 216, 193, 2)
12	Daughter	(1070–1135)	Gisele of Burgundy	(172, 216, 2, 115, 107)
13	Daughter	(1092–1154)	Adelaide de Savoy	(135, 172, 216, 224, 2)
14	Son	(1126–1183)	Peter I de Courtenay	(161, 163, 216, 221, 2)
15	Daughter	(1160–1218)	Alice de Courtenay	(216, 221, 2, 115, 107)
16	Daughter	(1188–1246)	Isabel Taillefer	(144, 195, 216, 2, 42)
17	Son	(1206–1272)	Henry III of England	(144, 171, 216, 2, 37)
18	Son	(1239–1307)	Edward I Plantagenet	(51, 161, 175, 171, 167)
19	Daughter	(1272–1307)	Joan of Acre	(51, 161, 175, 167, 216)
20	Daughter	(1292–1337)	Eleanor de Clare	(51, 163, 161, 2, 37)
21	Daughter	(1326–1389)	Elizabeth le Despenser	(146, 2, 42, 260, 287)
22	Son	(1352–1416)	Thomas IV de Berkeley	(146, 158, 2, 42, 260)

(continued)

Table C.21 (continued)

Gen	Relation	(Birth–Death)	Person	(Source numbers)
23	Daughter	(1387–1422)	Elizabeth de Berkeley	(146, 158, 2, 42, 260)
24	Daughter	(1408–1467)	Eleanor Beauchamp	(51, 144, 161, 165, 2)
25	Daughter	(1437–1501)	Eleanor Beaufort	(51, 144, 184, 2, 42)
26	Daughter	(1472–)	Margaret Spencer	(42, 259, 260, 293, 287)
27	Son	(1495–1529)	William Carey	(51, 144, 146, 161, 168)
28	Daughter	(1524–1568)	Mary Catherine Carey	(51, 161, 164, 143, 259)
29	Son	(1548–)	Henry Knolles	(158, 260, 589, 956)
30	Daughter	(1578–)	Lettice Knolles	(158, 589)
31	Son	(1609–1678)	William Paget	(161, 175, 158, 258, 589)
32	Daughter	(1643–)	Penelope Paget	(175, 158, 589)
33	Son	(1669–1739)	Paul Foley	(175, 589)
34	Daughter	(1708–1748)	Penelope Foley	(175, 589)
35	Daughter	(1740–1770)	Mary Howard	(589)
36	Son	(1766–1848)	Robert Waring Darwin	(589, 268)
37	Son	(1809–1882)	Charles Robert Darwin	(304, 589, 711, 1318, 1502)

Table C.22. Lineage of maximal ascent from Victor Emmanuel II of Sardinia to Charlemagne

Gen	Relation	(Birth–Death)	Person	(Source numbers)
0	Start	(747–814)	Charlemagne the Great	(51, 144, 150, 1, 2)
1	Son	(778–840)	Louis I the Pious	(51, 144, 150, 216, 2)
2	Son	(795–855)	Lothaire I of Italy	(144, 150, 175, 216, 193)
3	Son	(835–869)	Lothair II of Lorraine	(51, 150, 216, 30, 107)
4	Daughter	(852–907)	Gisela of Lorraine	(150, 216, 107, 328, 1689)
5	Daughter	(868–917)	Ragnhildis von Friesland	(150, 216, 107)
6	Daughter	(896–968)	Matilda von Ringleheim	(51, 150, 2, 115, 107)
7	Daughter	(913–984)	Gerberge of Saxony	(51, 144, 150, 216, 193)
8	Daughter	(943–981)	Mathilde of France	(150, 216, 2, 30, 107)
9	Daughter	(965–1017)	Gerberga of Upper Burgundy	(150, 194, 115, 107, 181)
10	Daughter	(990–1043)	Gisela of Swabia	(150, 115, 107, 293, 311)
11	Son	(1017–1056)	Henry III the Mild	(216, 150, 2, 106, 114)

(continued)

Table C.22. (continued)

Gen	Relation	(Birth–Death)	Person	(Source numbers)
12	Son	(1050–1106)	Henry IV of Holy Roman Empire	(150, 216, 2, 35, 96)
13	Daughter	(1073–1143)	Agnes of Franconia	(150, 2, 35, 106, 115)
14	Daughter	(1088–1110)	Hedwig-Eilike Hohenstaufen	(150, 244, 1376)
15	Daughter	(1103–1170)	Heilika von Lengenfeld-Hopfenohe	(150, 130, 244, 328, 566)
16	Daughter	(1130–1174)	Hedwig von Wittelsbach	(150, 244)
17	Son	(1159–1204)	Berthold VI of Meran	(216, 150, 115, 107, 130)
18	Daughter	(1181–1213)	Gertrude von Meran	(150, 216, 2, 95, 113)
19	Son	(1206–1270)	Bela IV of Hungary	(150, 216, 2, 62, 95)
20	Son	(1239–1272)	Stephen V of Hungary	(216, 2, 106, 113, 115)
21	Daughter	(1257–1323)	Marie Arpad of Hungary	(216, 2, 100, 106, 113)
22	Daughter	(1273–1299)	Marguerite de Naples	(216, 2, 69, 84, 90)
23	Daughter	(1293–1342)	Jeanne of Valois	(216, 2, 95, 115, 107)
24	Daughter	(1310–1356)	Marguerite of Hainault	(216, 150, 106, 260, 1393)
25	Son	(1336–1404)	Albert I of Bavaria	(150, 216, 224, 239, 206)
26	Daughter	(1363–1423)	Margaret of Bavaria	(150, 216, 52, 260, 311)
27	Daughter	(1407–1476)	Agnes of Burgundy	(216, 52, 311, 324, 258)
28	Daughter	(1436–1465)	Isabella of Bourbon	(216, 311, 324, 589, 1393)
29	Daughter	(1457–1482)	Mary of Burgundy	(150, 216, 62, 76, 104)
30	Son	(1478–1506)	Philip I of Austria	(150, 216, 194, 62, 95)
31	Son	(1503–1564)	Ferdinand I of Austria	(150, 206, 216, 63, 62)
32	Daughter	(1528–1590)	Anna of Austria	(150, 311, 589)
33	Daughter	(1551–1608)	Maria Anne of Bavaria	(150, 154, 113, 311, 589)
34	Daughter	(1584–1611)	Margareta of Austria	(150, 62, 113, 311, 589)
35	Daughter	(1606–1646)	Maria of Spain	(150, 52, 104, 113, 589)
36	Son	(1640–1705)	Leopold I Ignace of Austria	(150, 216, 1, 62, 104)
37	Son	(1678–1711)	Joseph I of Austria	(51, 150, 1, 63, 72)
38	Daughter	(1699–1757)	Maria Josepha of Austria	(150, 113, 258, 311, 589)
39	Daughter	(1724–1760)	Maria-Amalia of Saxony	(150, 216, 108, 311, 589)
40	Daughter	(1745–1792)	Marie Louise of Bourbon-Spain	(150, 216, 113, 108, 311)

(continued)

Table C.22. (continued)

Gen	Relation	(Birth–Death)	Person	(Source numbers)
41	Son	(1769–1824)	Ferdinand III of Habsburg-Lorraine	(78, 216, 150, 106, 311)
42	Daughter	(1801–1855)	Maria Theresa Francisca of Tuscany	(150, 216, 311, 258, 340)
43	Son	(1820–1878)	Victor Emmanuel II of Sardinia	(78, 150, 216, 96, 106)

Table C.23 Lineage of maximal ascent from James Clerk Maxwell to Charlemagne

Gen	Relation	(Birth–Death)	Person	(Source numbers)
0	Start	(747–814)	Charlemagne the Great	(51, 144, 150, 1, 2)
1	Son	(778–840)	Louis I the Pious	(51, 144, 150, 216, 2)
2	Daughter	(798–860)	Hildegarde Rotrude Adeltrude of France	(150, 938, 711, 107, 30)
3	Son	(815–857)	Erispoe II de Vannes	(1, 52, 95, 107, 1689)
4	Daughter	(830–880)	Lotitia de Poher	(328, 1154, 1689, 107)
5	Son	(860–888)	Berenger Judicael I of Rennes	(216, 328, 1689, 107)
6	Daughter	(883–925)	Papie de Bayeux	(216, 2, 95, 115, 107)
7	Son	(900–942)	William I Long Sword	(150, 175, 216, 2, 32)
8	Son	(933–996)	Richard I of Normandy	(175, 184, 171, 216, 2)
9	Son	(970–1027)	Richard II of Normandy	(51, 150, 161, 216, 2)
10	Son	(1002–1035)	Robert I of Normandy	(51, 158, 161, 164, 166)
11	Daughter	(1030–1084)	Adelaide of Normandy	(164, 173, 216, 2, 52)
12	Daughter	(1054–1086)	Judith of Lens	(216, 2, 52, 58, 95)
13	Daughter	(1073–1131)	Matilda of Huntingdon	(216, 2, 52, 73, 95)
14	Son	(1114–1152)	Henry of Huntingdon	(158, 171, 166, 216, 2)
15	Son	(1144–1219)	David of Huntingdon	(155, 158, 161, 175, 164)
16	Daughter	(1206–1251)	Isabella of Huntingdon	(158, 166, 216, 9, 42)
17	Son	(1224–1295)	Robert Bruce	(158, 166, 216, 9, 42)
18	Son	(1243–1304)	Robert David Bruce	(158, 166, 172, 170, 216)
19	Son	(1274–1329)	Robert I Bruce	(158, 164, 175, 179, 173)
20	Daughter	(1296–1316)	Marjorie Bruce	(158, 179, 216, 9, 42)

(continued)

Table C.23 (continued)

Gen	Relation	(Birth–Death)	Person	(Source numbers)
21	Son	(1316–1390)	Robert II Stewart	(144, 158, 175, 172, 162)
22	Daughter	(1362–1446)	Elizabeth Stewart	(166, 216, 42, 240, 244)
23	Daughter	(1381–)	Elizabeth Lindsay	(240, 259, 287, 258, 589)
24	Son	(1410–1493)	Thomas Erskine	(167, 184, 173, 170, 258)
25	Son	(1446–1509)	Alexander Erskine	(184, 170, 258, 589)
26	Son	(1474–1513)	Robert Erskine	(170, 173, 184, 258, 589)
27	Daughter	(1510–1572)	Catherine Erskine	(184, 589, 956, 1376)
28	Daughter	(1533–)	Marjory Elphinstone	(184, 956)
29	Son	(1553–1610)	John Drummond	(956)
30	Son	(1585–1649)	William Drummond	(956)
31	Daughter	(1630–)	Elizabeth Drummond	(956)
32	Daughter	(1650–)	Elizabeth Henderson	(956)
33	Son	(1676–1755)	John Clerk	(956)
34	Son	(1715–1784)	George Clerk	(956)
35	Son	(1750–1793)	James Clerk	(956)
36	Son	(1790–1856)	John Clerk Maxwell	(956)
37	Son	(1831–1879)	James Clerk Maxwell	(1318, 1502, 956)

Table C.24 Lineage of maximal ascent from Thomas Alva Edison to Charlemagne

Gen	Relation	(Birth–Death)	Person	(Source numbers)
0	Start	(747–814)	Charlemagne the Great	(51, 144, 150, 1, 2)
1	Son	(778–840)	Louis I the Pious	(51, 144, 150, 216, 2)
2	Son	(795–855)	Lothaire I of Italy	(144, 150, 175, 216, 193)
3	Son	(835–869)	Lothair II of Lorraine	(51, 150, 216, 30, 107)
4	Daughter	(852–907)	Gisela of Lorraine	(150, 216, 107, 328, 1689)
5	Daughter	(868–917)	Ragnhildis von Friesland	(150, 216, 107)
6	Daughter	(896–968)	Matilda von Ringleheim	(51, 150, 2, 115, 107)
7	Daughter	(913–984)	Gerberge of Saxony	(51, 144, 150, 216, 193)
8	Daughter	(930–973)	Aubree of Lorraine	(216, 193, 2, 115, 107)
9	Daughter	(950–1005)	Ermentrude de Roucy	(216, 2, 115, 107, 315)
10	Son	(990–1057)	Raynalt I of Burgundy	(51, 161, 216, 2, 95)

(continued)

Table C.24 (continued)

Gen	Relation	(Birth–Death)	Person	(Source numbers)
11	Son	(1024–1087)	William I of Burgundy	(144, 172, 216, 193, 2)
12	Daughter	(1062–1102)	Matilda of Burgundy	(216, 2, 115, 107, 315)
13	Daughter	(1080–1142)	Helie of Burgundy	(216, 2, 115, 107, 227)
14	Daughter	(1118–1174)	Ela of Ponthieu	(216, 2, 260, 293, 258)
15	Daughter	(1137–1199)	Isabel de Warenne	(155, 216, 2, 73, 95)
16	Daughter	(1164–)	Isabel de Warenne	(260, 293)
17	Son	(1191–1225)	Hugh II Le Bigod	(2, 42, 80, 37, 260)
18	Daughter	(1212–)	Isabel Le Bigod	(165, 2, 42, 260, 287)
19	Daughter	(1240–1301)	Maud FitzJohn	(167, 172, 2, 42, 260)
20	Daughter	(1266–1306)	Isabel de Beauchamp	(51, 161, 165, 172, 216)
21	Daughter	(1282–1322)	Maud de Chaworth	(216, 2, 42, 149, 260)
22	Daughter	(1320–1362)	Mary of Lancaster	(144, 173, 216, 42, 66)
23	Son	(1341–1407)	Henry de Percy	(51, 144, 158, 161, 164)
24	Daughter	(1368–)	Margaret de Percy	(144, 260, 711)
25	Son	(1401–1459)	Henry Fenwick	(260, 711)
26	Daughter	(1430–)	Ann Fenwick	(711)
27	Daughter	(1462–1520)	Elizabeth Radcliffe	(711)
28	Daughter	(1497–)	Winifred Threlkeld	(711)
29	Daughter	(1528–1573)	Jane Pickering	(711)
30	Daughter	(1549–1617)	Constance Hamby	(711)
31	Daughter	(1580–1662)	Elizabeth Marsh	(711, 6)
32	Daughter	(1619–1680)	Mary Thomas	(711, 6)
33	Daughter	(1646–1721)	Deliverance Hall	(1146)
34	Daughter	(1690–1723)	Patience Durfee	(1146)
35	Son	(1710–1803)	Benjamin Tallman	(1146)
36	Daughter	(1746–)	Mercy Tallman	(1146, 711)
37	Daughter	(1783–1850)	Mercy Peckham	(711)
38	Daughter	(1810–1871)	Nancy Matthews Elliott	(240, 711, 234)
39	Son	(1847–1931)	Thomas Alva Edison	(240, 711, 1318, 1502, 234)

Table C.25 Lineage of maximal ascent from Max Karl Planck to Charlemagne

Gen	Relation	(Birth–Death)	Person	(Source numbers)
0	Start	(747–814)	Charlemagne the Great	(51, 144, 150, 1, 2)
1	Son	(778–840)	Louis I the Pious	(51, 144, 150, 216, 2)
2	Son	(795–855)	Lothaire I of Italy	(144, 150, 175, 216, 193)
3	Son	(835–869)	Lothair II of Lorraine	(51, 150, 216, 30, 107)
4	Daughter	(852–907)	Gisela of Lorraine	(150, 216, 107, 328, 1689)
5	Daughter	(868–917)	Ragnhildis von Friesland	(150, 216, 107)
6	Daughter	(896–968)	Matilda von Ringleheim	(51, 150, 2, 115, 107)
7	Daughter	(913–984)	Gerberge of Saxony	(51, 144, 150, 216, 193)
8	Daughter	(943–981)	Mathilde of France	(150, 216, 2, 30, 107)
9	Daughter	(965–1017)	Gerberga of Upper Burgundy	(150, 194, 115, 107, 181)
10	Daughter	(990–1043)	Gisela of Swabia	(150, 115, 107, 293, 311)
11	Son	(1017–1056)	Henry III the Mild	(216, 150, 2, 106, 114)
12	Son	(1050–1106)	Henry IV of Holy Roman Empire	(150, 216, 2, 35, 96)
13	Daughter	(1073–1143)	Agnes of Franconia	(150, 2, 35, 106, 115)
14	Daughter	(1088–1110)	Hedwig-Eilike Hohenstaufen	(150, 244, 1376)
15	Daughter	(1103–1170)	Heilika von Lengenfeld-Hopfenohe	(150, 130, 244, 328, 566)
16	Daughter	(1130–1174)	Hedwig von Wittelsbach	(150, 244)
17	Son	(1159–1204)	Berthold VI of Meran	(216, 150, 115, 107, 130)
18	Daughter	(1181–1213)	Gertrude von Meran	(150, 216, 2, 95, 113)
19	Daughter	(1207–1231)	Elizabeth of Hungary	(150, 216, 95, 130, 2)
20	Daughter	(1224–1275)	Sophia of Thuringia	(144, 150, 193, 109, 107)
21	Son	(1244–1308)	Henry I of Hesse	(51, 144, 150, 193, 212)
22	Daughter	(1268–1317)	Adelheid of Hesse	(193, 224, 130, 1393, 109)
23	Son	(1299–1347)	Heinrich VIII of Henneberg-Schleusingen	(193, 204, 150, 224, 311)
24	Daughter	(1319–1384)	Elisabeth of Henneberg	(204, 224, 193, 1393)
25	Son	(1341–1388)	Ulrich of Wurttemberg	(193, 150, 106)
26	Son	(1363–1417)	Eberhard III of Wurttemberg	(150, 193, 268, 106)
27	Son	(1388–1419)	Eberhard IV of Wurttemberg	(193, 200, 199, 224, 106)

(continued)

Table C.25 (continued)

Gen	Relation	(Birth–Death)	Person	(Source numbers)
28	Daughter	(1418–1447)	Elisabeth von Dagersheim	(711)
29	Daughter	(1442–1490)	Elizabeth Antonia Lyher	(711)
30	Son	(1472–1537)	Philipp Volland	(711)
31	Son	(1505–)	(son) Volland	(711)
32	Son	(1530–)	Michael Volland	(711)
33	Daughter	(1574–)	Susanne Volland	(711)
34	Son	(1606–1660)	Johann Konrad Lang	(711)
35	Son	(1657–1738)	Conrad Philipp Lang	(711)
36	Son	(1688–1763)	Gottlieb Christian I Lang	(711)
37	Daughter	(1732–1799)	Veronika Dorothea Lang	(711)
38	Son	(1751–1833)	Gottlieb Jakov Planck	(711)
39	Son	(1785–1831)	Heinrich Ludwig Planck	(711)
40	Son	(1817–1900)	Johann Julius Wilhelm Planck	(711)
41	Son	(1858–1947)	Max Karl Ernst Ludwig Planck	(1318, 1502, 711)

Table C.26 Lineage of maximal ascent from Henry Ford to Charlemagne

Gen	Relation	(Birth–Death)	Person	(Source numbers)
0	Start	(747–814)	Charlemagne the Great	(51, 144, 150, 1, 2)
1	Son	(778–840)	Louis I the Pious	(51, 144, 150, 216, 2)
2	Son	(795–855)	Lothaire I of Italy	(144, 150, 175, 216, 193)
3	Son	(835–869)	Lothair II of Lorraine	(51, 150, 216, 30, 107)
4	Daughter	(852–907)	Gisela of Lorraine	(150, 216, 107, 328, 1689)
5	Daughter	(868–917)	Ragnhildis von Friesland	(150, 216, 107)
6	Daughter	(896–968)	Matilda von Ringleheim	(51, 150, 2, 115, 107)
7	Daughter	(913–984)	Gerberge of Saxony	(51, 144, 150, 216, 193)

(continued)

Table C.26 (continued)

Gen	Relation	(Birth–Death)	Person	(Source numbers)
8	Daughter	(930–973)	Aubree of Lorraine	(216, 193, 2, 115, 107)
9	Daughter	(950–1005)	Ermentrude de Roucy	(216, 2, 115, 107, 315)
10	Daughter	(995–1068)	Agnes of Burgundy	(150, 216, 2, 115, 107)
11	Son	(1022–1058)	William VII of Aquitaine	(216, 224, 193, 206, 2)
12	Daughter	(1058–1129)	Clementia of Poitiers	(216, 224, 193, 206, 115)
13	Daughter	(1088–1137)	Yolande of Wassemberg	(216, 206, 2, 115, 107)
14	Son	(1108–1171)	Baldwin IV of Hainault	(135, 216, 224, 2, 106)
15	Son	(1150–1195)	Baldwin V of Hainault	(3, 216, 2, 106, 115)
16	Daughter	(1170–1190)	Isabelle of Hainault	(216, 2, 52, 84, 113)
17	Son	(1187–1226)	Louis VIII of France	(135, 216, 2, 52, 66)
18	Son	(1215–1270)	Louis IX of France	(216, 193, 2, 52, 66)
19	Son	(1245–1285)	Philip III the Bold	(51, 216, 193, 2, 52)
20	Son	(1270–1325)	Charles III de Valois	(138, 161, 216, 224, 221)
21	Daughter	(1293–1342)	Jeanne of Valois	(216, 2, 95, 115, 107)
22	Daughter	(1312–1369)	Philippa of Hainault	(165, 216, 2, 42, 52)
23	Son	(1340–1399)	John of Gaunt	(138, 51, 144, 146, 155)
24	Son	(1372–1410)	John I Beaufort	(51, 161, 216, 2, 42)
25	Son	(1406–1455)	Edmund I Beaufort	(51, 144, 184, 2, 42)
26	Daughter	(1437–1501)	Eleanor Beaufort	(51, 144, 184, 2, 42)
27	Daughter	(1477–1542)	Katherine Spencer	(144, 42, 260, 287, 311)
28	Daughter	(1505–1547)	Eleanor Percy	(1376)
29	Daughter	(1525–1600)	Alice Neville	(1146)
30	Son	(1556–1631)	Ralph Josselyn	(1146)
31	Son	(1591–1661)	Thomas Josselyn	(1146)
32	Son	(1619–1670)	Abraham Josselyn	(1146)
33	Son	(1652–1730)	Henry Josselyn	(1146)
34	Son	(1697–)	Henry Josselyn	(1146)
35	Daughter	(1741–1806)	Lucy Josselyn	(1376)
36	Son	(1775–1831)	William Ford	(1376)
37	Son	(1799–1842)	Jonathan Ford	(1376, 956)
38	Son	(1826–1905)	William Ford	(1376)
39	Son	(1863–1947)	Henry Ford	(1318, 1502, 711)

Table C.27 Lineage of maximal ascent from Wilbur Wright to Charlemagne

Gen	Relation	(Birth–Death)	Person	(Source numbers)
0	Start	(747–814)	Charlemagne the Great	(51, 144, 150, 1, 2)
1	Son	(778–840)	Louis I the Pious	(51, 144, 150, 216, 2)
2	Son	(795–855)	Lothaire I of Italy	(144, 150, 175, 216, 193)
3	Son	(835–869)	Lothair II of Lorraine	(51, 150, 216, 30, 107)
4	Daughter	(852–907)	Gisela of Lorraine	(150, 216, 107, 328, 1689)
5	Daughter	(868–917)	Ragnhildis von Friesland	(150, 216, 107)
6	Daughter	(896–968)	Matilda von Ringleheim	(51, 150, 2, 115, 107)
7	Daughter	(913–984)	Gerberge of Saxony	(51, 144, 150, 216, 193)
8	Daughter	(930–973)	Aubree of Lorraine	(216, 193, 2, 115, 107)
9	Daughter	(950–1005)	Ermentrude de Roucy	(216, 2, 115, 107, 315)
10	Son	(990–1057)	Raynalt I of Burgundy	(51, 161, 216, 2, 95)
11	Son	(1024–1087)	William I of Burgundy	(144, 172, 216, 193, 2)
12	Daughter	(1062–1102)	Matilda of Burgundy	(216, 2, 115, 107, 315)
13	Daughter	(1080–1142)	Helie of Burgundy	(216, 2, 115, 107, 227)
14	Daughter	(1118–1174)	Ela of Ponthieu	(216, 2, 260, 293, 258)
15	Daughter	(1137–1199)	Isabel de Warenne	(155, 216, 2, 73, 95)
16	Daughter	(1164–)	Isabel de Warenne	(260, 293)
17	Son	(1191–1225)	Hugh II Le Bigod	(2, 42, 80, 37, 260)
18	Daughter	(1212–)	Isabel Le Bigod	(165, 2, 42, 260, 287)
19	Daughter	(1240–1301)	Maud FitzJohn	(167, 172, 2, 42, 260)
20	Daughter	(1266–1306)	Isabel de Beauchamp	(51, 161, 165, 172, 216)
21	Daughter	(1282–1322)	Maud de Chaworth	(216, 2, 42, 149, 260)
22	Daughter	(1311–1372)	Eleanor of Lancaster	(184, 216, 2, 42, 112)
23	Son	(1340–1369)	Henry II de Beaumont	(51, 161, 184, 42, 260)
24	Son	(1361–1396)	John II de Beaumont	(165, 184, 168, 42, 80)
25	Daughter	(1389–1488)	Elizabeth de Beaumont	(168, 165, 234, 260, 258)
26	Daughter	(1406–1477)	Margaret Botreaux	(158, 168, 165, 184, 172)
27	Daughter	(1426–)	Alice Hungerford	(234)
28	Son	(1456–1512)	Robert II White	(260, 258, 1146, 234)
29	Son	(1475–)	Robert III White	(260, 1146, 234)
30	Son	(1495–1549)	Thomas White	(144, 247, 1146, 234)
31	Son	(1516–1578)	Richard White	(247, 260, 1146, 234)
32	Son	(1542–1600)	Robert IV White	(247, 1146, 234)

(continued)

Table C.27 (continued)

Gen	Relation	(Birth–Death)	Person	(Source numbers)
33	Son	(1561–1617)	Robert V White	(247, 259, 293, 575, 1146)
34	Daughter	(1600–1647)	Anna Rosanna White	(247, 67, 259, 293, 589)
35	Son	(1635–1689)	Samuel II Porter	(67, 240, 247, 259, 656)
36	Son	(1676–1750)	Experience Porter	(240, 247, 1376)
37	Son	(1704–1772)	John Porter	(240, 247, 1376)
38	Daughter	(1734–1777)	Sarah Porter	(240, 247, 1376)
39	Daughter	(1762–1848)	Sarah Freeman	(240, 247, 1376)
40	Son	(1790–1861)	Daniel Wright	(240, 247, 711)
41	Son	(1828–1917)	Milton Wright	(240, 247, 711)
42	Son	(1867–1912)	Wilbur Wright	(240, 247, 711, 1318, 1502)

Appendix D: Sources Referenced in the Lineages

The 180 source identifiers listed after the persons in Appendices B and C are shown. For each source, the number of persons referenced by the source within the Research Genealogy is listed after the tag #inGen. For sources published in Geneanet, the total number of persons identified by the source is also listed.

1S: Book, Gardner, Laurence, *Bloodline of the Holy Grail*, Barnes & Noble, #inGen 985.

2S: Book, Jordan, Wilfred, *Colonial and Revolutionary Families of Pennsylvania*, Lewis historical Publishing Company, Inc., 1952, Vol 13, Library of Congress, F148.C72, #inGen 3953.

3S: Book, Norwich, John Julius, *A Short History of Byzantium*, Alfred A. Knopf, #inGen 553.

6S: Web Site, Latter Day Saints Genealogy Database, https://www.familysearch.org, Change Date 17 OCT 2020, #inGen 684.

9S: Book, Davies, John, *A History of Wales*, Allen Lane, The Penquin Press, 1993, #inGen 246.

15S: Book, Baigent, Michael, Richard Leigh, Henry Lincoln, *Holy Blood, Holy Grail*, Dell Book, 1982, #inGen 189.

22S: CD ROM, World family Tree, Vol v702-01, #inGen 1577.

28S: Book, Evans, Gwynfor, *Wales a History, 2000 years of Welsh History*, Barnes and Noble, 1996, #inGen 45.

30S: Book, McKitterick, Rosamond, *The Frankish Kingdoms under the Carolingians*, Longman, 1983, DC 70.M3 1983, #inGen 296.

31S. Book, Nelson, Janet I., *Charles the Bald*, Longman, 1992, DC 76.N45 1992, #inGen 107.

32S: Book, Sturluson, Snorri, *Heimskringla, History of the Kings of Norway*, Lee M. Hollandeer, University of Texas Press, Austin, 1964, #inGen 617.

35S: Book, Archer, *English History from Contemporary Writers*, F. York Powell, G. P. Putnam's Sons, 1889, D 163.A87, The Crusade of Richard I, #inGen 144.

37S: Book, Moody, T. W., F. X. Martin, F. J. Byrne, *A New History of Ireland*, Oxford Clarendon Press, 1984, Vol IX, DA 912.N48, #inGen 2156.

© The Editor(s) (if applicable) and The Author(s), under exclusive license to Springer Nature Switzerland AG 2023
R. W. Moore, *Trustworthy Communications and Complete Genealogies*, Synthesis Lectures on Information Concepts, Retrieval, and Services,
https://doi.org/10.1007/978-3-031-16836-9

42S: Book, Weis, Frederic Lewis, Arthur Adams, *The Magna Charta Sureties*, 1215, Walter Lee Sheppard, Jr., Genealogical Publishing Co., Inc., 1982, #inGen 4463.

46S: Book, Wurts, John S., *Magna Charta*, Brookfield Publishing Co., Philadelphia 1954, CS 419.W82, #inGen 124.

48S: Book, Ryan, Joseph F., *The Russian Chronicles, a Thousand Years that Changed the World*, CLB, Quadrillion Publishing Ltd, 1998, #inGen 127.

49S: Periodical, Bayne, W. Wilfrid, De Ayala of Castile, Augustan Society Magazine, Vol XIII:6, 289–291, #inGen 68.

51S: Book, Brydges, Egerton, *Collins's Peerage of England*, AMS Press 1970, #inGen 4575.

52S: Book, of Albany, Prince Michael, *The Forgotten Monarchy of Scotland*, Element Books Inc., Boston 1998, 1590, #inGen 1573.

58S: Book, Cockayne, George E., *The Complete Peerage*, The St. Catherine Press Ltd, London, 1910, Vol I, #inGen 15.

59S: Book, Riasanovsky, Nicholas V., *A History of Russia*, Oxford University Press, 1963, #inGen 96.

62S: Book, Berenger, Jean, *A History of the Habsburg Empire 1273–1700*, Longman, New York, 1994, #inGen 129.

63S: Book, Wandruszka, Adam, *The House of Habsburg, Six Hundred Years of a European Dynasty*, Doubleday, New York, 1964, #inGen 207.

66S: chart, Bennett, Archibald F., Genealogical Society of Utah, The Kinship of Families, #inGen 851.

67S: Book, *Burke's Presidential Families of the United States of America*, Burke's Peerage Limited, London, 1981, Vol 2nd Edition, CS 69.B82 1981, #inGen 1964.

69S: Book, Muir, D. Erskine, *Machiavelli and His Times*, William Heinemann Ltd, London, 1936, #inGen 44.

71S: Book, Fay, Sidney Bradshaw, *The Rise of Brandenburg-Prussia to 1786*, Henry Holt and Company, New York, 1937, #inGen 35.

72S: Book, Kroll, Maria, *Sophie, Electress of Hanover*, Victor Gollancz LTD, London, 1973, #inGen 111.

73S: Book, *Kings and Queens*, Diamond Books, 1994, Diagram Visual Information Limited, #inGen 805.

76S: Book, 1998 *Grolier Multimedia Encyclopedia*, Grolier Interactive Inc., #inGen 387.

78S: Book, McNaughton, Arnold, *The Book of Kings*, Quadrangle, The New York Times Book Company, #inGen 6847.

80S: Book, Curtis, Edmund, *A History of Medieval Ireland from 1086 to 1513*, Barnes & Noble, Inc., 1968, DA 933.C8 1968, #inGen 531.

81S: Book, Collins, Roger, *The Arab Conquest of Spain 710–797*, Basil Blackwell Inc. 1989, DP99.C58, #inGen 17.

84S: Book, Fawtier, Robert, *The Capetian Kings of France*, MacMillan & Co Ltd 1960, DC 82 F313, #inGen 65.

90S: Book, Deviosse, Jean, *Jean le Bon*, Fayard 1985, DC 99.D48 1985, #inGen 83.

95S: Book, Ashley, Mike, *The Mammoth Book of British Kings and Queens*, Carroll & Graf Publishers, Inc. New York, 1998, #inGen 2542.

96S: Book, Katz, Robert, *The Fall of the House of Savoy*, The Macmillan Company, 1971, DG 611.5 K36, #inGen 87.

100S: Book, Cotterill, H. B., *Italy from Dante to Tasso (1300–1600)*, George G. Harrap & Company Ltd., 1919, DG 533 C6 1919, #inGen 120.

102S: Book, Chiappini, Luciano, *Gli Estensi*, dall'Oglio, 1967, DG 463.8 E7 C5, #inGen 285.

104S: Book, Payne, Stanley G., *A History of Spain and Portugal*, University of Wisconsin Press, 1973, DP66 P382, #inGen 216.

106S: Book, Morby, John E., *Dynasties of the World*, Oxford University Press, 1989, 929.7 Mor, #inGen 2529.

107S: Book, Stuart, Roderick W., *Royalty for Commoners*, Genealogical Publishing Company, 1998, 929.7 Stu, #inGen 4027.

108S: Book, Williamson, David, *Debrett's Kings and Queens of Europe*, Salem House Publishers, 1988, 929.7 Wil, #inGen 1620.

109S: Web Site, http://members.aol.com/_ht_a/dwidad/toc.html, 9/6/00, #inGen 2228.

112S: Web Site, http://www.geocities.com/Athens/Academy/3677/d0.htm, 9/7/00, #inGen 273.

113S: chart, Taute, Anne, University of North Carolina Press, Kings and Queens of Europe, 1989, #inGen 1240.

114S: Book, Alen, Rupert and Anna Marie Dahlquist, *Royal Families of Medieval Scandinavia, Flanders, and Kiev*, Kings River Publications, #inGen 688.

115S: Chart, Churchyard, James Nohl, http://freepages.rootsweb.com/ ~jonnic/genealogy/People/zUnknownConnections/Churchyard/genealgy.html, Churchyard/Orr Family Museum, Change Date 17 OCT 2020, #inGen 853.

117S: Web Site, http://www.schlossgarten.de/personen/herrsher/herzoege/html, #inGen 5.

119S: Web Site, https://www.ancestry.com/search/collections/1030/, Ancestry, Change Date 17 OCT 2020, #inGen 75.

129S: Web Site, Brandyberry, Edward L., Descendants of Burchard of Zollern (Brandyberry Line), http://www.familytreemaker.com/users/b/r/a/Edward-L-Brandyberry/GENE2-0017.html, 1/7/2001, #inGen 29.

130S: Web Site, Lindwall, Bo, King Erik XIV's ancestors in 14 generations, http://www.genealogi.se/erikeng.htm, 1/7/2001, #inGen 1247.

132S: Book, Carl Stephenson, *Mediaeval History, Europe from the second to the Sixteenth Century*, Harper & Row, 1962, D 117 S8 1962, #inGen 63.

135S: Book, Isenburg, Wilhelm, *Stammtafeln zur Geswchichte der Europaischen Staaten*, Vol 1.3, CS616.I722, Change Date 17 OCT 2020, #inGen 455.

138S: Book, Isenburg, Wilhelm, *Stammtafeln zur Geswchichte der Europaischen Staaten*, Vol I.2, CS616.I722, Change Date 17 OCT 2020, #inGen 91.

143S: Book, Wagner, Anthony Richard, *English Genealogy*, Oxford at the Clarendon Press, 1972, CS 414.W3 1972, #inGen 464.

144S: Book, Collins, Arthur, *Collins's Peerage of England*, AMS Press, 1970, Vol 2, CS 421.C74 1970 V.2, Change Date 2 MAY 2020, #inGen 4121.

146S: Book, Collins, Arthur, *Collins's Peerage of England*, AMS Edition, 1970, Vol 3, CS 421.C74 1970 v.3, #inGen 5935.

149S: Book, Fraser, Antonia, *The Lives of the Kings and Queens of England*, Alfred A. Knopf, New York, 1975, #inGen 407.

150S: Book, Schwennicke, Detlev, *Europaische Stammtafeln, Neue Folge*, Vittorio Klostermann, Frankfurt am Main, 1998, Vol I.1, Change Date 17 OCT 2020, #inGen 10232.

152S: Book, Giffard, Ambrose Hardinge, *Who was my Grandfather*, Harrison and Sons, London, 1865, CS 439.GA34 1865, #inGen 116.

153S: Book, Weis, F. L., *Ancestral Roots of Sixty Colonists*, Vol 6th Edition, Baltimore, 1988, #inGen 44.

154S: Book, Schwennicke, Detlev, *Europaische Stammtafeln*, Marburg, 1989, Vol III.3, Change Date 17 OCT 2020, #inGen 204.

155S: Book, Brydges, Egerton, *Collins's Peerage of England*, Collins, Arthur, AMS Press, Inc., 1970, Vol IV, CS 42.C74 1970, Change Date 2 MAY 2020, #inGen 4980.

158S: Book, Brydges, Egerton, *Collins's Peerage of England*, Collins, Arthur, AMS Press Inc., New York, 1970, Vol 5, CS 421.C74 1970 V.5, #inGen 6418.

160S: Book, Burke, Bernard, *Burke's Guide to the Royal Family: Burke's genealogical series, Burke's Peerage, 1973*, Change Date 17 OCT 2020, #inGen 468.

161S: Book, Brydges, Sir Egerton, *Collins's Peerage of England*; Genealogical, Biographical and Historical, AMS Press, Inc., Vol 6, CS 421.C74 1970, #inGen 6756.

162S: Book, Cockayne, G. E., *The Complete Peerage*, Sutton Publishing, Vol 9, Change Date 17 OCT 2020, #inGen 1068.

163S: Book, Cockayne, G. E., *The Complete Peerage*, Sutton Publishing, Vol 4, Change Date 17 OCT 2020, #inGen 1372.

164S: Book, Cockayne, G. E., *The Complete Peerage of England, Scotland, Ireland, Great Britain, and the United Kingdom*, Sutton Publishing, Vol 1, Change Date 17 OCT 2020, #inGen 1211.

165S: Book, Cockayne, G. E., *The Complete Peerage*, Sutton Publishing, Vol 2, Change Date 17 OCT 2020, #inGen 1237.

166S: Book, Cockayne, G. E., *The Complete Peerage*, Sutton Publishing, Vol 3, Change Date 17 OCT 2020, #inGen 1197.

167S: Book, Cockayne, G. E., *The Complete Peerage*, Sutton Publishing, Vol 5, Change Date 17 OCT 2020, #inGen 1482.

168S: Book, Cockayne, G. E., *The Complete Peerage*, Sutton Publishing, Vol 6, Change Date 17 OCT 2020, #inGen 812.

169S: Book, Cockayne, G. E., *The Complete Peerage*, Sutton Publishing, Vol 7, Change Date 17 OCT 2020, #inGen 1087.

170S: Book, Cockayne, G. E., *The Complete Peerage*, Sutton Publishing, Vol 8, Change Date 17 OCT 2020, #inGen 1037.

171S: Book, Cockayne, G. E., *The Complete Peerage*, Sutton Publishing, Vol 10, Change Date 17 OCT 2020, #inGen 987.

172S: Book, Cockayne, G. E., *The Complete Peerage*, Sutton Publishing, Vol 11, Change Date 17 OCT 2020, #inGen 1250.

173S: Book, Cockayne, G. E., *The Complete Peerage*, Sutton Publishing, Vol 12, Change Date 17 OCT 2020, #inGen 1889.

175S: Book, Brydges, Sir Egerton, *Collins's Peerage of England*, AMS Press Inc., Vol 7, CS 421.C74 1970, #inGen 6256.

179S: Book, Brydges, Egerton, *Collins's Peerage of England*; Genealogical, Biographical, and Historical, AMS Press Inc., 1970, Vol 8, #inGen 7733.

181S: Book, Riche, Pierre, *The Carolingians, A Family Who Forged Europe*, Allen, Michael Idomir, University of Pennsylvania Press, #inGen 327.

184S: Book, Brydges, Sir Egerton, *Collins's Peerage of England*, AMS Press, Inc., Vol 9, #inGen 6314.

193S: Book, Schwennicke, Detlev, *Europaische Stammtafeln, Neu Folge*, Vittorio Klostermann, Vol I.2, Change Date 17 OCT 2020, #inGen 7842.

194S: Book, Schwennicke, Detlev, *Europaische Stammtafeln, Neue Folge*, Vittorio Klostermann, Frankfurt am Main, Vol I.3, Change Date 17 OCT 2020, #inGen 7732.

195S: Book, Schwennicke, Detlev, *Europaische Stammtafeln*, Verlag von J. A. Stargardt, 1989, Vol 3.4, Change Date 17 OCT 2020, #inGen 582.

199S: Book, Schwennicke, Detlev, *Europaische Stammtafeln*, Verlag von J. A. Stargardt, 1986, Vol 11, Change Date 17 OCT 2020, #inGen 382.

200S: Book, Schwennicke, Detlev, *Europaische Stammtafeln*, Verlag von J. A. Stargardt, 1992, Vol 12, Change Date 17 OCT 2020, #inGen 788.

203S: Web Site, Vans, Jamie, Vans Family Archive, http://homepages.rootsweb.anc estry.com/~vfarch/genealogy-data/wc04/wc04_067.html, Change Date 18 MAR 2013, #inGen 29.

204S: Book, Schwennicke, Detlev, *Europaische Stammtafeln*, Verlag von J. A. Stargardt, 1995, Vol 16, Change Date 17 OCT 2020, #inGen 752.

205S: Book, Schwennicke, Detlev, *Europaische Stammtafeln*, Vittorio Klostermann, 1998, Vol 17, Change Date 17 OCT 2020, #inGen 1427.

206S: Book, Schwennicke, Detlev, Europaische Stammtafeln, Vittorio Klostermann, 1998, Vol 18, Change Date 17 OCT 2020, #inGen 5171.

212S: Book, Schwennicke, Detlev, *Europaische Stammtafeln*, Verlag von J. A. Stargardt, Vol 4, Change Date 17 OCT 2020, #inGen 265.

216S: Book, Schwennicke, Detlev, *Europaische Stammtafeln, Stammtafeln zur Geschichte der Europaischen Staaten*, Verlag von J. A. Stargardt, Vol 2, Change Date 17 OCT 2020, #inGen 1161.

221S: Book, Schwennicke, Detlev, *Europaische Stammtafeln*, Verlag von J. A. Stargardt, Marburg, 1979, Vol Book VII, CS 616 E8 1978 v. 7, Change Date 17 OCT 2020, #inGen 1017.

223S: Web Site, Dowling, Tim, Diana Frances Spencer: Family Tree, https://gw.gen eanet.org/tdowling, Geneanet.org, Change Date 18 OCT 2020, #inGen 204.

224S: Book, Isenburg, Wilhelm, *Europaische Stammtafeln*, Verlag von J. A. Stargardt, 1956, Vol 3.1, CS 616 I7 1956 v. 3, Change Date 17 OCT 2020, #inGen 14,696.

226S: Web Site, Stirnet Genealogy, http://www.stirnet.com/HTML/genie/genfam.htm, 8/18/05, #inGen 109.

227S: Web Site, Bradley, Hal, Bradley, Collette, Gillespie & Opp Ancestry, http://fre epages.genealogy.rootsweb.com/~hwbradley/, 8/16/05, #inGen 3415.

234S: Web Site, Weebers, Hans A. M., Genealogy of the Presidents of the USA, http:// users.legacyfamilytree.com/USPresidents/6415.htm, Oct 11, 2005, #inGen 1917.

239S: Book, Schwennicke, Detlev, *Europaische Stammtafeln*, Verlag von J. A. Stargardt, Vol III.2, 1983, Change Date 17 OCT 2020, #inGen 8137.

240S: Web Site, Reitwiesner, William Addams, http://www.wargs.com/political/bush. html, Ancestry of George W. Bush, June 24, 2006, #inGen 8566.

244S: Web Site, Reitwiesner, William Addams, William Addams Reitwiesner Genealogical Services, http://www.wargs.com/, Change Date 4 APR 2008, #inGen 7652.

247S: Web Site, Lawson, Stephen M., Some Notable Cousins, http://kinnexions.com/ kinnexions/cousins.htm, Aug 14, 2006, #inGen 9590.

258S: Book, Roberts, Gary Boyd, *The Royal Descents of 600 Immigrants to the American Colonies or the United States*, Genealogical Publishing Company, Incorporated, #inGen 12816.

259S: Book, Roberts, Gary Boyd, *Ancestors of American Presidents*, New England Historical Genealogical Society; Boston Massachusetts, 1995, #inGen 6416.

260S: Book, Richardson, Douglas, *Plantagenet Ancestry*, Everingham, Kimball G., Genealogical Publishing Company, 2004, #inGen 17909.

262S: Book, Rixford, Elizabeth M. Leach, Families Directly Descended from All the Royal Families in Europe (495–1923) and Mayflower Descendants, Genealogical Publishing Company, Inc, 2002, #inGen 314.

268S: Web Site, Wikipedia, Wikipedia, The Free Encyclopedia, http://en.wikipedia. org/wiki/, September 23, 2007, #inGen 612.

270S: Web Site, Castelli, Jorge H., Tudor Place, http://www.tudorplace.com.ar/WEN TWORTH.htm, Oct 5, 2007, #inGen 124.

287S: Book, Weis, Frederick Lewis, *The Magna Charta Sureties*, 1215, Walter Lee Sheppard, Jr., William R. Beall, Genealogical Publishing Co., Inc, 5th, Change Date 12 MAY 2008, #inGen 3091.

293S: Book, Hart, Craig, *A Genealogy of the Wives of the American Presidents and Their First Two Generations of Descent*, McFarland & Company, Inc, Change Date 6 OCT 2008, #inGen 5858.

303S: Book, Pfafman, Robert Frederick, *Ancestry and Progeny of Captain James Blount, Immigrant*, GR 929.2 B657P, Change Date 3 JAN 2009, #inGen 3308.

304S: Book, Palmer, R.R. and Joel Colton, *A History of the Modern World*, Alfred A Knopf, Change Date 25 JAN 2009, #inGen 604.

307S: Web Site, Sherman, Don, *The Genealogy of the Honorable Roger Sherman*, http://web.archive.org/web/20041029024521/http://members.aol.com/macpinhead/soyarticle04.html, 21 Feb 2009, Change Date 21 FEB 2009, #inGen 107.

311S: Book, von Redlich, Marcellus Donald R., *Pedigrees of some of the Emperor Charlemagne's Descendants*, Genealogical Publishing Co., Vol 1, Change Date 16 MAY 2009, #inGen 4326.

315S: Book, Turton, W. H., *The Plantagenet Ancestry*, Genealogical Publishing Co., Inc., Change Date 12 SEP 2009, #inGen 653.

324S: Web Site, Velde, Francois, *The French Royal Family: A Genealogy*, http://www.heraldica.org/topics/france/roygenea.htm, Change Date 20 FEB 2010, #inGen 1119.

328S: Book, Koman, Alan J., *A Who's Who of your Ancestral Saints*, Genealogical Publishing Company, Change Date 12 MAR 2010, #inGen 1170.

333S: Web Site, Castelli, Jorge, Welcome to my Tudor Court, http://www.tudorplace.com.ar, Change Date 3 JUL 2010, #inGen 320.

334S: Web Site, Lundy, Darryl, the Peerage, http://thepeerage.com, 3 Jul 2010, Change Date 17 OCT 2020, #inGen 338.

340S: Book, Williams, George, *Papal Genealogy, the Families and Descendants of the Popes*, McFarland & Company, Change Date 6 SEP 2010, #inGen 2019.

342S: Web Site, Camp, J., Papal Cousins, http://freepages.genealogy.rootsweb.ancestry.com/~cbjbar/chartpopes.html, Change Date 6 SEP 2010, #inGen 313.

343S: Web Site, La Corte, John Cilia, Genealogy and Heraldry, http://cilialacorte.com/index.html, Change Date 18 SEP 2010, #inGen 2601.

382S: Book, Starr, Brian Daniel, *Lineage of the Saints*, BookSurge Publishing, (North Charleston, South Carolina; 2007), https://books.google.com, Change Date 18 OCT 2020, #inGen 92.

414S: Book, Madariaga, Salvador, *Bolivar*, Shocken Books (New York, 1952), #inGen 20.

415S: Web Site, Tantagente, Tantagente family, http://freepages.genealogy.rootsweb.ancestry.com/~tantagente/fam/fam00194.html, 25 Oct 2010, #inGen 372.

448S: Web Site, Lewis, Marlyn, Our Royal, Titled, Noble, and Commoner Ancestors & Cousins, http://our-royal-titled-noble-and-commoner-ancestors.com, Change Date 13 NOV 2010, #inGen 626.

493S: Web Site, Wikipedia, The Free Encyclopedia, http://en.wikipedia.org/wiki/, Change Date 4 DEC 2010, #inGen 1531.

501S: Book, Johnston, George Harvey, *The Heraldry of the Douglases: With notes on all the males of the family, descriptions of the arms, plates and pedigrees*, W. & A. K. Johnston, Limited, http://books.google.com/books?id=FwwtAAAAYAAJ, Change Date 5 DEC 2010, #inGen 4.

502S: Web Site, The Douglas Archives, http://www.douglashistory.co.uk/famgen/index.php, Change Date 5 DEC 2010, #inGen 27.

503S: Book, Ross, Ian Simpson, *The Life of Adam Smith*, Google Books, Change Date 5 DEC 2010, #inGen 6.

504S: Book, Johnston, George Harvey, *The Heraldry of the Douglases*, Google books, Change Date 5 DEC 2010, #inGen 2.

507S: Book, Pogue, Kate Emery, *Shakespeare's Family*, Praeger, Change Date 5 DEC 2010, #inGen 27.

566S: Web Site, Wikipedia, http://wikipedia.org/wiki/, 09/20/10, Online public encyclopedia, #inGen 1514.

575S: Web Site, Matthiesen, Diana Gale, My Family's Genealogy and Related Projects, http://dgmweb.net/GenealogyHome.html, Change Date 11 FEB 2011, #inGen 518.

589S: Web Site, Harrison, Bruce, The Family Forest Descendants of King Edward III of England and Queen Philippa of Hainault, http://www.familyforest.com, Change Date 8 JAN 2018, #inGen 30,315.

643S: Web Site, Whitney Research Group, Whitney Research Group, http://wiki.whitneygen.org/wrg/index.php/Main_Page, Change Date 4 FEB 2012, #inGen 280.

651S: Web Site, FamousKin, FamousKin.com, http://famouskin.com/lewg04.aspx#4961, Change Date 9 FEB 2012, #inGen 122.

656S: Web Site, Marshall, John, Hooker Family, http://homepages.rootsweb.ancestry.com/~marshall/esmd284.htm, Change Date 18 FEB 2012, #inGen 91.

664S: Web Site, Findagrave, Find a Grave, http://www.findagrave.com/cgi-bin/fg.cgi?page=gr&GRid = 24738573, Change Date 25 JAN 2014, #inGen 970.

711S: Web Site, Geni, Geni, http://www.geni.com/, Change Date 24 DEC 2012, #inGen 47144.

714S: Web Site, *Gray, Wally and Frances, Sharing our Links to the Past*, http://www.geocities.com/~wallyg/godiva.htm, 16 Aug 2012, Change Date 16 AUG 2012, #inGen 48.

724S: Web Site, Familypedia, http://familypedia.wikia.com/wiki/Thomas_Bolling_(1735-1804), Change Date 11 OCT 2012, #inGen 33.

763S: Web Site, Barney, Eliza Starbuck, Barney Genealogical Record, http://www.nantuckethistoricalassociation.net/bgr/BGR-o/index.htm, Change Date 4 FEB 2013, #inGen 299.

852S: Web Site, Allen, Jamie, Ancestors of Prince William of England, http://fabped igree.com/willa1.htm, Change Date 1 SEP 2013, #inGen 1522.

860S: Web Site, Hughes, David, 5 descent-lines of Queen Elizabeth of Britain from Julia, Sister of Julius Caesar, http://www.angelfire.com/ego/et_deo/empire2britain.wps. htm, Change Date 21 OCT 2013, #inGen 689.

936S: Web Site, Wikipedia, Wikipedia, http://it.wikipedia.org/wiki/, Change Date 8 JUN 2014, #inGen 361.

938S: Web Site, Mozzoni, Cicogna, *Genealogia Cicogna Mozzoni e famiglie corre-late*, http://gw.geneanet.org/fcicogna, 8 Jun 2014, 190,313 persons, Change Date 17 MAR 2022, #inGen 1059.

956S: Web Site, Wikitree, Wikitree, http://www.wikitree.com, Change Date 24 JAN 2016, #inGen 9910.

964S: Web Site, Allen, Jamie, FabPed Genealogy Vers. 74, http://fabpedigree.com/s052/f781599.htm, Change Date 10 APR 2016, #inGen 242.

1137S: Web Site, van den Brempt, Hein, Family tree, http://gw.geneanet.org/heinvdb_f, Change Date 4 NOV 2016, 629,124 persons, #inGen 1108.

1146S: Web Site, Dowling, Tim, Tim Dowling's Family Tree, http://gw.geneanet.org/tdowling, Change Date 9 NOV 2016, 794,315 persons, #inGen 12523.

1149S: Web Site, Mattos e Silva, Jose and Antonio, Association of the Histori-cal Nobility of Portugal, http://www.socgeografialisboa.pt/wp/wp-content/uploads/2010/01/MAGALHÃES-19–11-2010.pdf, Change Date 17 NOV 2016, #inGen 12.

1154S: Web Site, Geneanet, Pierfit, http://gw.geneanet.org/pierfit, Change Date 18 OCT 2020, #inGen 1651.

1155S: Web Site, Illig, Remy, Geneanet, http://gw.geneanet.org/illig, Change Date 18 OCT 2020, #inGen 125.

1162S: Web Site, Webtrees, Bislew Family, http://bislew.com/family_tree/family.php?famid=F53&ged=Bislew%20Family, Change Date 19 DEC 2016, #inGen 15.

1163S: Web Site, Gontard, Jean, Jean Gontard's Family Tree, http://gw.geneanet.org/jeangontard, Change Date 19 DEC 2016, 93,161 persons, #inGen 14.

1191S: Web Site, Polier, Christoph, Christoph Graf von Polier's Family Tree, http://gw.geneanet.org/cvpolier, Change Date 4 FEB 2017, 242589 persons, #inGen 1243.

1205S: Web Site, Willems, Kees, Geneanet, http://gw.geneanet.org/kewi01, Change Date 18 OCT 2020, 392,721 persons, #inGen 103.

1251S: Web Site, Terlinden, Jean-Charles, l'Ascendance Terlinden-de Potesta's Family Tree, http://gw.geneanet.org/lard, Change Date 18 JUL 2020, 150,889 persons, #inGen 114.

1263S: Web Site, Herlyn, Menno, Webtrees, Herlin Family, http://familie-herlyn.de/webtrees/individual.php?pid=I13198&ged=familie_herlyn.ged, Change Date 4 JAN 2018, #inGen 44.

1281S: Web Site, Dardenne, Henry, 149,608 persons, *Base des families Dardenne, Thibaut, Debailleul, Jeroon, & Gobert*, https://gw.geneanet.org/dardhen, 18 Jun 2018, Change Date 18 APR 2022, #inGen 28.

1318S: chart, CoreGen3 Research Genealogy, Notable Persons, Change Date 24 MAR 2019, #inGen 2369.

1376S: Web Site, FamilySearch, FamilySearch, https://ancestors.familysearch.org, Change Date 26 NOV 2019, #inGen 24911.

1393S: Book, Hansen, Charles M., Thompson, Neil D., *The Ancestry of Charles II, King of England, A Medieval Heritage*, McNaughton & Gunn, Inc., Change Date 27 MAR 2020, #inGen 1594.

1407S: Web Site, Taverney, Olivier, Olivier Taverney's Family Tree, https://gw.geneanet.org/geneta, Change Date 6 APR 2020, 88976 persons, #inGen 43.

1472S: Web Site, Myheritage, Myheritage, https://www.myheritage.com, Change Date 14 AUG 2020, #inGen 30.

1502S: Book, Hart, Michael, *The 100: A Ranking of the Most Influential Persons in History*, 1 Jun 2000, Carol Publishing Group, Change Date 18 JUL 2021, #inGen 88.

1517S: Web Site, Wikipedia, *Merovingian dynasty*, https://en.wikipedia.org/wiki/Merovingian_dynasty, 27 Aug 2021, Change Date 27 AUG 2021, #inGen 198.

1520S: Web Site, Wikipedia, Yngling Dynasty, https://en.wikipedia.org/wiki/Yngling, Change Date 4 SEP 2021, #inGen 78.

1548S: Web Site, Uken, Gunter, Gunter Uken's Family Tree, https://gw.geneanet.org/guken2, Change Date 24 NOV 2021, 372642 persons, #inGen 538.

1556S: Web Site, Gallais, Andre, 34,556 persons, *Clodion*, https://gw.geneanet.org/agallais, Change Date 15 APR 2022, #inGen 137.

1564S: Web Site, Cauchy, Alain, Alain Cauchy, https://gw.geneanet.org/alain3131, Change Date 30 NOV 2021, 59457 persons, #inGen 392.

1589S: Web Site, Poe, Clarissa, *Clarissa Poe's Family Tree*, https://gw.geneanet.org/frozenpark, 96,897 persons, Change Date 16 FEB 2022, #inGen 15.

1641S: Web Site, Dufour, Alain, *Alain Dufour*, https://gw.geneanet.org/alaindufour11, 124,457 persons, Change Date 24 FEB 2022, #inGen 313.

1642S: Web Site, Wikipedia, *Kings of the Visigoths family Tree*, https://en.wikipedia.org/wiki/Visigothic_Kingdom#List_of_kings, Change Date 26 FEB 2022, #inGen 88.

1660S: Web Site, Besson, Michel, *Michel Besson's Family Tree*, https://gw.geneanet.org/michelbesson, 156,115 persons, Change Date 17 MAR 2022, #inGen 89.

1664S: Web Site, Grojean, Claude, *Bienvenue sur l'arbre de Claude Grojean & Colette Ginet*, https://gw.geneanet.org/cgrojean, 193,199 persons, Change Date 22 MAR 2022, #inGen 107.

1680S: Web Site, Guerin, Thierry, *Genealogie G SdM*, https://gw.geneanet.org/thigue, 7,984 persons, Change Date 28 MAR 2022, #inGen 446.

1683S: Web Site, gratienne, *gratienne's Family Tree*, https://gw.geneanet.org/gratienne, 30,942 persons, Change Date 15 APR 2022, #inGen 820.

1689S: Web Site, Chatelain, Stephane, Compiled by Claude Chatelain, Historian, *Stephane Chatelain's Family Tree*, https://gw.geneanet.org/citron06, 32,614 persons, Change Date 8 MAY 2022, #inGen 2314.

1709S: Web Site, Juillie, Yves, *Yves Juillie's Family Tree*, https://gw.geneanet.org/juillie, 69, I701 persons, Change Date 20 MAY 2022, #inGen 35.

Appendix E: Algorithms Used to Analyze Genealogies

Genealogy graph-traversal algorithms normally are used to identify the ancestors of a "root" person, or the descendants, or the relatives. Collectively the ancestors, descendants, and relatives form an extended family of the "root" person.

In the development of a Genealogical History of the Modern World, additional graph-traversal algorithms were used to explore lineage coalescence. All the analyses listed below can be applied by the CoreGen3 genealogy analysis workbench to any genealogy that is exported as a Gedcom file. The analyses can be used to identify Progenitors for persons of Western European descent. The corresponding CoreGen3 commands are listed for each algorithm. The analyses require the identification of a root person (for ancestors), an ancestral person (for descendants and lineages) and a core person (for the root of the unifying ancestry).

1. Common ancestors of a group of persons. Select Group Op, then the group, then click on Common Ancestors.
   a. The ancestors of the first member of the group are identified and marked.
   b. For each additional member of the group, the ancestors are found. For each marked ancestor, a check is made that they are included in the ancestors for each additional group member. If not, the person is unmarked.
   c. The result is a list of common ancestors of a group of persons.
2. Progenitors for the common ancestors of a group of persons. Select Group Op, then the group, then click on Common Ancestors.
   a. The common ancestors of the group of persons are found.
   b. The most recent common ancestor is found, and the person is marked as a progenitor.
   c. The ancestors of the marked person and the marked person are deleted from the set of common ancestors.
   d. The process is repeated until no one is left in the list.
   e. This creates a set of progenitors for the group. The progenitors and their ancestors are the common ancestors of the group of persons.

© The Editor(s) (if applicable) and The Author(s), under exclusive license to Springer Nature Switzerland AG 2023
R. W. Moore, *Trustworthy Communications and Complete Genealogies*, Synthesis Lectures on Information Concepts, Retrieval, and Services,
https://doi.org/10.1007/978-3-031-16836-9

3. Number of ascents to each ancestor from a root person. Each ascent is a unique path through the genealogy starting from the root person to an ancestor. Select Root Op, then click on Ascents.
   a. The ancestors of the root person are identified.
   b. The number of ascents to an ancestor is then set to the sum of the number of ascents to each child of the ancestor.
4. Number of descents to each descendant from an ancestral person. Each descent is a unique path through the genealogy starting from the ancestral person to a descendant. Select Root Op, then click on Descents.
   a. The descendants of the root person are identified.
   b. The number of descents to a descendant is then set to the sum of the number of descents to each parent of the descendant.
5. Partitioning of the genealogy into extended families. Select Root Op, then click on Partition.
   a. All the ancestors, descendants, and relatives of the root person are found.
   b. A list of all unrelated spouses of the descendants and relatives is generated.
   c. For each unrelated spouse, the members of their extended family, excluding persons already in an extended family, are marked with a unique partition number.
   d. When the list has been processed, this generates a ring of extended families around the original partition.
   e. Generate a new list of all unrelated spouses from the last ring of extended families and iterate.
   f. This produces successive rings of extended families about the original partition.
   g. At the end of the process, all persons in the genealogy should be a member of an extended family. Persons without a partition number are islands disconnected from the rest of the genealogy.
6. Descendant fractional distribution. Select Root Op, then click on DescDNA.
   a. The descendants of the starting ancestral person are identified.
   b. The descendant fractional distribution of the starting ancestral person is set to 1.
   c. The descendant fractional distribution of each child is set to half the sum of the descendant fractional distributions of the parents.
7. Ancestral fractional distribution. Select Root Op, then click on AncDNA.
   a. The ancestors of the starting root person are identified.
   b. The ancestral fractional distribution of the root person is set to 1.
   c. The ancestral fractional distribution of each parent is set to half the sum of the ancestral fractional distributions of their children.
8. Global connectivity metric. Select Core Op, then the Core person, then click on Core Own Cousin.
   a. The number of ascents to each ancestor from a root person are identified.
   b. Global connectivity metric = Log2(number of ascents) / (number of generations)
9. Own cousin relationship. Select Root Op, then click on OwnCous.

    a. The ancestors of the root person are identified.

    b. For each ancestor of the root person, the closest ancestor with two ascents is identified.

    c. The own cousin relationship is set to the number of generations to the closest ancestor minus 1.

    d. A counter for the closest ancestor with two ascents is incremented to track the number of times that person is the lineage coalescence point.

10. Local connectivity metric. Select Core Op, then the Core person, then click on Core Own Cousin.

    a. The ancestors of the starting root person are identified.

    b. The own cousin relationship is generated for each ancestor.

    c. For each lineage coalescence point, the number of generations from the ancestor is added and a counter for the number of times is incremented, #Conn.

    d. The average number of generations from an ancestor to the lineage coalescence point is generated, #Gen.

    e. The local connectivity metric is set to Log2(#Conn)/#Gen.

11. Lineage of maximal ascent. Select Root Op, then click on Max Ascents.

    a. The ancestors of the root person are identified.

    b. The number of ascents is generated for each ancestor.

    c. Starting with the ancestral person, the child with the largest number of ascents is selected.

    d. The process iterates back to the root person.

12. Lineage of maximal descent. Select Root Op, then click on Descents.

    a. The descendants of a starting ancestral person are identified.

    b. The number of descents is generated for each descendant.

    c. Starting with a root person, the parent with the largest number of descents is selected.

    d. The process iterates back up to the ancestral person.

13. Unifying Ancestry. Select Group Op, choose Royal Lines, then click on Common Ancestors.

    a. The common ancestors of the Kings of Belgium, Great Britain, the Netherlands, Norway, Spain, and Sweden and the Queen of Denmark, are identified.

    b. A descendant of one of the Kings and Queens is selected as the root person for the Unifying Ancestry.

    c. The ancestors of the root person are identified. This automatically includes the common ancestors of the Royal Lines.

14. Connection year to Unifying Ancestry. Select Core Op, choose Core person, then click on Connection Year.

    a. Identify the relatives of the root person for the Unifying Ancestry.

    b. For each relative, identify the birth year of the closest common ancestor with the root person.

15. Progenitors for the Unifying Ancestry. Select Root Op, then click on Progenitors.
    a. The descendants and relatives of the Ancestral person are identified.
    b. The members of the Unifying Ancestry are identified.
    c. The number of ascents to each member of the Unifying Ancestry is calculated.
    d. The own cousin relationship is generated for each member of the Unifying Ancestry.
    e. Progenitors are selected that are:
       i.   Members of the Unifying Ancestry.
       ii. Persons who are the lineage coalescence points for own-cousin relationships
       iii. Persons who are not an ancestor of the Ancestral person
       iv. Persons who have more than 200 million ascents from the Root person.
       v.   Persons born after 200 AD
       vi. Persons that have descendants alive after 1900
       vii. Persons with at least 700 more descendants than the number of descendants held in common with the ancestral person
       viii. Persons that have at least the number of generations from the Root person defined by the missing information analysis
       ix. Persons that are not ancestors of another progenitor
16. Missing information analysis. Select Core Op, choose Core person, then click on Core Own Cousin.
    a. The own cousin relationship is generated for each member of the Unifying Ancestry
    b. The candidates are restricted to members of the common ancestors of the Royal Families of Western Europe
    c. The average number of generations to the closest lineage coalescence point is calculated, g.
    d. The average number of lineages that coalesce is calculated across all lineage coalescence points, c.
    e. The number of ascents is modeled as $c^{(n/g)}$.
    f. The number of ascents to a progenitor should be at least the initial population size.
    g. If $g/\log2(c)$ is greater than 1.77, no essential information is missing for identifying progenitors.
    h. The minimum number of generations to a progenitor is set to $g / \log2(c) * \log2(\text{population-size})$
    i. The best unifying ancestry has the smallest value for the minimum number of generations to a progenitor
17. Genealogical History of the Modern World. Select Root Op, then click on Sources. Search for source 304 (Palmer) or source 1502 (Hart) or source 1318 (notable persons).
    a. The Unifying Ancestry for persons of Western European Descendants is selected.
    b. The notable persons in the selected source are identified.

c. For each notable person, a lineage linking the person to the Unifying Ancestry is identified within the Research Genealogy.

d. This identifies familial relationships between persons involved in historical events.

e. The history of the modern world can now be interpreted through interactions between relatives. My favorite is that Isabella I of Castile (who financed Christopher Columbus) is the Grandniece of Henry the Navigator (who financed voyages along the coast of Africa).

Glossary

Ancestral fractional distribution Measures the fraction of the ancestry that the root person receives from each ancestor.

Ancestral person Person to which a lineage is traced from the root person.

Ascent A unique path from a root person to an ancestor.

Authenticity Information that ensures the preserved document has not been altered.

Authoritative source Directly documents an historical event. Examples are birth certificates, marriage certificates, nobility patents.

Closure Property that measures whether information elements are isolated, disconnected from the rest of the collection.

Coherence Property that measures whether information elements can be linked to external knowledge bases (external relationships), and whether information elements needed to justify conclusions about the unifying concept are missing (internal relationships).

Common ancestors The persons who claim as descendants all members of a group.

Connection year The birth year of the person who is the common ancestor of the root person and a relative.

Completeness Property that measures whether information elements can be linked to a unifying concept.

Connectivity Property that measures how the information elements are linked.

Consistency Property that measures whether all the attributes needed to interpret an information element have been defined.

Context The information needed to understand the content of communications.

Core child A child that is an ancestor of the root person for the unifying ancestry.

Correctness Property that measures whether all information values fall within acceptable ranges.

Degrees of Separation Number of extended families that must be traversed to link a person to the root person.

Descendant fractional distribution Measures the fraction of the ancestry of each descendant that comes from the ancestral person.

© The Editor(s) (if applicable) and The Author(s), under exclusive license
to Springer Nature Switzerland AG 2023
R. W. Moore, *Trustworthy Communications and Complete Genealogies*, Synthesis Lectures on Information Concepts, Retrieval, and Services,
https://doi.org/10.1007/978-3-031-16836-9

Descent A unique path from an ancestral person to a descendant.

Description Information that defines the meaning of the record.

Essential information metric Identifies the number of generations needed to trace lineages to a progenitor, based on a requirement that the average number of ascents to a progenitor is at least the population size.

Extended family Consists of the spouses, ancestors, descendants, and relatives of a person.

Global connectivity metric Divides Log2(number of ascents) by the number of generations to an ancestor.

Integrity Information that verifies the document has not been corrupted.

Knowledge base The set of relationships that define how information elements can be organized, such as spatial, structural, temporal, procedural, functional, algorithmic, logical, semantic, systemic, and epistemological relationships.

Lineage A path through the genealogy that links the root person to a relative.

Lineage coalescence The merging of two lineages to a common ancestor through two children that are ancestors of a root person.

Lineage of eldest descent The linage found by selecting the eldest child who is an ancestor of the root person and iterating down to the root person.

Lineage of maximal ascent The lineage found by starting with the ancestral person, selecting the child with the largest number of ascents, and iterating back to the root person.

Lineage of maximal descent The lineage found by starting with the root person, selecting the parent with the largest number of descents, and iterating back to the ancestral person.

Lineage of youngest descent The lineage found by selecting the youngest child who is an ancestor of the root person and iterating down to the root person.

Local connectivity metric Divides Log2(number of times a person is an own cousin lineage coalescence point) by the average number of generations to the person.

Log2(x) Logarithm to base 2, which counts the number of factors of 2 present in x.

Own cousin relationship The number of generations minus 1 to the closest ancestor with two ascents.

Own cousin lineage coalescence point The closest ancestor with two ascents.

Partition a genealogy Identify an extended family for the root person, then generate extended families for each unrelated spouse, and iterate until everyone is a member of an extended family.

Preservation The management of authenticity, integrity, representation, description, and provenance information for each document.

Progenitor Ancestor from whom descents can be traced to each member of a group of persons, such as U. S. Presidents, or Western Europeans.

Provenance Information that identifies the creator of the document.

Representation Information that defines how to interpret the format of a document.

Root person Person whose ancestors are determined.

Spousal relationship The cousin relationship between a husband and a wife.

Treetop An ancestor that has either zero or one parent.

Treetop of own cousin lineage coalescence point An ancestor that is an own cousin lineage coalescence point but does not have an own cousin relationship.

Unifying Ancestry The ancestors of a root person to which all members of a national community should be able to link their ancestry.

Printed in the United States
by Baker & Taylor Publisher Services

Printed in the United States
by Baker & Taylor Publisher Services